Lecture Notes in Computer Science   11668

Commenced Publication in 1973
Founding and Former Series Editors:
Gerhard Goos, Juris Hartmanis, and Jan van Leeuwen

FoLLI Publications on Logic, Language and Information
Subline of Lectures Notes in Computer Science

More information about this series at http://www.springer.com/series/7407

Raffaella Bernardi · Greg Kobele ·
Sylvain Pogodalla (Eds.)

Formal Grammar

24th International Conference, FG 2019
Riga, Latvia, August 11, 2019
Proceedings

 Springer

Editors
Raffaella Bernardi
University of Trento
Povo, Italy

Greg Kobele
Universität Leipzig
Leipzig, Germany

Sylvain Pogodalla
LORIA, Inria Nancy
Villers-lès-Nancy, France

ISSN 0302-9743   ISSN 1611-3349   (electronic)
Lecture Notes in Computer Science
ISBN 978-3-662-59647-0   ISBN 978-3-662-59648-7   (eBook)
https://doi.org/10.1007/978-3-662-59648-7

LNCS Sublibrary: SL1 – Theoretical Computer Science and General Issues

This Springer imprint is published by the registered company Springer-Verlag GmbH, DE
part of Springer Nature
The registered company address is: Heidelberger Platz 3, 14197 Berlin, Germany

Preface

The Formal Grammar conference series (FG) provides a forum for the presentation of new and original research on formal grammar, mathematical linguistics, and the application of formal and mathematical methods to the study of natural language. Themes of interest include, but are not limited to:

- Formal and computational phonology, morphology, syntax, semantics, and pragmatics
- Model-theoretic and proof-theoretic methods in linguistics
- Logical aspects of linguistic structure
- Constraint-based and resource-sensitive approaches to grammar
- Learnability of formal grammar
- Integration of stochastic and symbolic models of grammar
- Foundational, methodological, and architectural issues in grammar and linguistics
- Mathematical foundations of statistical approaches to linguistic analysis

Previous FG meetings were held in Barcelona (1995), Prague (1996), Aix-en-Provence (1997), Saarbrücken (1998), Utrecht (1999), Helsinki (2001), Trento (2002), Vienna (2003), Nancy (2004), Edinburgh (2005), Malaga (2006), Dublin (2007), Hamburg (2008), Bordeaux (2009), Copenhagen (2010), Ljubljana (2011), Opole (2012), Düsseldorf (2013), Tübingen (2014), Barcelona (2015), Bolzano-Bozen (2016), Toulouse (2017), and Sofia (2018).

FG 2019, the 24th conference on Formal Grammar, was held in Riga, Latvia, on August 11, 2019. The conference comprised an invited talk, by Tal Linzen, and seven contributed papers selected from 11 submissions. The present volume includes an abstract of the invited talk and the contributed papers.

We would like to thank the people who made the 24th FG conference possible: the invited speaker, the members of the Program Committee, and the organizers of ESSLLI 2019, with which the conference was colocated.

August 2019

Raffaella Bernardi
Greg Kobele
Sylvain Pogodalla

Organization

Program Committee

Berthold Crysmann	CNRS - LLF (UMR 7110), Paris-Diderot, France
Philippe de Groote	Inria Nancy – Grand Est, France
Nissim Francez	Technion - IIT, Israel
Thomas Graf	Stony Brook University, USA
Laura Kallmeyer	Heinrich-Heine-Universität Düsseldorf, Germany
Makoto Kanazawa	Hosei University, Japan
Stepan Kuznetsov	Steklov Mathematical Institute, Russian Federation
Alessandro Lenci	University of Pisa, Italy
Robert Levine	Ohio State University, USA
Glyn Morrill	Universitat Politècnica de Catalunya, Spain
Reinhard Muskens	Tilburg Center for Logic and Philosophy of Science, The Netherlands
Stefan Müller	Freie Universität Berlin, Germany
Mark-Jan Nederhof	University of St Andrews, UK
Rainer Osswald	Heinrich-Heine-Universität Düsseldorf, Germany
Christian Retoré	Université de Montpellier and LIRMM-CNRS, France
Mehrnoosh Sadrzadeh	Queen Mary University of London, UK
Manfred Sailer	Goethe University Frankfurt, Germany
Edward Stabler	UCLA and Nuance Communications, USA
Jesse Tseng	CNRS, France
Oriol Valentín	Universitat Politècnica de Catalunya, France
Christian Wurm	Heinrich-Heine-Universität Düsseldorf, Germany
Ryo Yoshinaka	Tohoku University, Japan

Standing Committee

Raffaella Bernardi	University of Trento, Italy
Greg Kobele	Universität Leipzig, Germany
Sylvain Pogodalla	Inria Nancy – Grand Est, France

What Inductive Biases Enable Human-Like Syntactic Generalization? (Abstract of Invited Talk)

Tal Linzen

Departments of Cognitive Science and Computer Science,
Johns Hopkins University, Baltimore, Maryland, USA
tal.linzen@jhu.edu
http://tallinzen.net/

Humans generalize their knowledge of language in a systematic way to syntactic structures that are rare or absent in their linguistic input, an observation traditionally discussed under the banner of "the poverty of the stimulus". This generalization behavior has motivated structural (innate) inductive biases. In this talk, I will demonstrate how neural networks with and without explicit syntactic structure can be used to test for the necessity and sufficiency of such structural biases. Focusing on subject-verb agreement as well as subject-auxiliary inversion in English question formation, I will argue that explicit structural biases are still necessary for human-like generalization in a learner trained on text only.

Contents

A Purely Surface-Oriented Approach to Handling Arabic Morphology

Yousuf Aboamer[(✉)] and Marcus Kracht

Bielefeld University, Postfach 10 10 31, 33501 Bielefeld, Germany
{yousuf.aboamer,marcus.kracht}@uni-bielefeld.de

Abstract. In this paper, we introduce a completely lexicalist approach to deal with Arabic morphology. This purely surface-oriented treatment is part of a comprehensive mathematical approach to integrate Arabic syntax and semantics using overt morphological features in the string-to-meaning translation. The basic motivation of our approach is to combine semantic representations with formal descriptions of morphological units. That is, the lexicon is a collection of signs; each sign δ is a triple $\delta = \langle E, C, M \rangle$, such that E is the exponent, C is the combinatorics and M is the meaning of the sign. Here, we are only concerned with the exponents, i.e. the components of a morphosemantic lexicon (for a fragment of Arabic). To remain surface-oriented, we allow for discontinuity in the constituents; constituents are sequences of strings, which can only be concatenated or duplicated, but no rule can delete, add or modify any string. Arabic morphology is very well known for its complexity and richness. The word formation in Arabic poses real challenges because words are derived from roots, which bear the core meaning of their derivatives, formed by inserting vowels and maybe other consonants. The units in the sequences are so-called glued strings rather than only strings. A glued string is a string that has left and right context conditions. Optimally morphs are combined in a definite and non-exceptional linear way, as in many cases in different languages (e.g. plural in English). The process of Arabic word formation is rather complex; it is not just a sequential concatenation of morphs by placing them next to each other. But the constituents are discontinuous. Vowels and more consonants are inserted between, before and after the root consonants resulting in what we call "fractured glued string", i.e. as a sequence of glued strings combined in diverse ways; forward concatenation, backward concatenation, forward wrapping, reduction, forward transfixation and, going beyond the multi-context free grammars (MCFGs), also reduplication.

Keywords: Discontinuity · Arabic morphology · Surface orientation · Morphosemantics

1 Introduction

Arabic is currently the most spoken Semitic language in the world with a number of speakers approaching 415 million, according to the CIA world factbook [1],

© Springer-Verlag GmbH Germany, part of Springer Nature 2019
R. Bernardi et al. (Eds.): FG 2019, LNCS 11668, pp. 1–17, 2019.
https://doi.org/10.1007/978-3-662-59648-7_1

and is the/an official language of 22 countries in a vast area extending from the Arabian Peninsula to the Fertile Crescent, in addition to some other countries in the middle east. It has recently become the focus of an increasing number of projects in natural language processing and computational linguistics [2]. In this regard, we investigate a new approach to integrate Arabic morphology, syntax and semantics. The central claim of this approach is that the argument structure provides an interface between syntax and semantics. The main function of the argument structure is to declare how the functor's semantic arguments are realized on the surface. [3][1] In this paper, we are only concerned with morphology, i.e. we show how to deal with Arabic morphology in a way that allows us to go beyond mere word formation and provide a fully compositional treatment of entire sentences. Sentences are composed from units that are combined in some way. Each unit is either a string, i.e. a sequence of letters, or a sequence of strings. Furthermore, to implement adjacency constraints we use so-called glued strings in place of strings, or sequences thereof. In order to deal with the complexity of Arabic morphology in a completely surface-oriented approach, we have to deal with a number of problems, especially the discontinuity of morphemes, matching among morphemes, and morphophonemic alternations resulting from the interaction among different types of morphs.

This paper consists of four parts in addition to the introduction. In Sect. 2, we discuss briefly the morphological system of Arabic and the various paradigms proposed for the study of Arabic word formation showing how the proposed approach differs from these paradigms. In Sect. 3 we give a description of the proposed approach and in Sect. 4 we show how we handle the morphology of Arabic within the proposed framework. Results and future work are indicated in Sect. 5.

2 Arabic Morphology: Theoretical and Computational Approaches

Arabic morphology is very well known for its complexity and richness. However, it exhibits rigorous and elegant logic [4]. It differs from that of English or other Indo-European languages because it is, to a large extent, based on discontinuous morphemes. Words in Arabic are derived from roots, which bear the core meaning of their derivatives, by inserting vowels and maybe other consonants. Roots are relatively invariable, discontinuous bound morphemes, typically containing three consonants in a certain order, and interlocked with vowels and other consonants to form stems [4]. Let us consider the following example: the triliteral (3 consonantal) root /د ر س/ ['d r s'] in Buckwalter's transliteration model,[2] is supposed to bear the meaning of studying and from which words

---

[1] The basic notions of the proposed approach are introduced in [3]. However, we concentrate in this paper on the most related ones to the nature of Arabic morphology and how they are applied to it.

[2] In this work, we use Buckwalter's transliteration model, but we made some modifications to make our work easier.

like /درس/ ['daras', 'he studied'] or ['durisa', 'be studied'] and /دارس/ ['dAris', 'student'] are derived. This process has evolved extensively and very productively in order to cover a vast array of meanings associated with each semantic field. The vast majority of Arabic words are derived in this way and therefore can be analyzed as consisting of two different bound morphs: a **root** and a so-called **pattern**, which interlock to form a word and neither of them can occur alone in the sentence. Moreover, after composing stems from discontinuous roots and some vowels or other consonants, they may concatenate with affixes or clitics, as clearly stated, for example, in the Arabic masculine sound plural form /معلمون/ ['muEalolimUna', 'teachers'] obtained from the concatenation of the strings / معلم / ['muEalolim', 'teacher'] and /ون/ ['Una', a nominal suffix for the masculine plural and a verbal suffix for masculine plural in indicative mood]. Stems may host suffixes that come after the stem as in this example, prefixes that come before the stem as in /يكتب/ ['yakotub', 'he writes'] obtained from /كتب/ ['kotub'] and /ي/ ['ya', third person singular of imperfective form] or both suffixes and prefixes as in /يكتبون/ ['yakotubUna', 'they write'] obtained from /كتب/ ['kotub'], /ي/ ['ya', third person singular of the imperfective form] and /ون/ ['Una', a suffix for the nominative masculine plural] [2].

As a result of such degree of richness and the complexity of word formation, Arabic morphology has been the focus of research in natural language processing for a long time [2, 5–9]. Researchers have adopted different approaches in the treatment of Arabic morphology, both theoretically and computationally [7]. Most of the efforts have been particularly devoted to addressing morphological analysis, generation and disambiguation. Arabic morphotactics have been sufficiently described and handled using finite state operations [10–13]. Al-Sughaiyer and Al-Kharashi [5] provide a comprehensive survey for Arabic morphological analysis techniques. Also, Dichy and Farghaly [7] count several approaches such as root and pattern, stem-based, machine learning among others. Depending on one of these approaches, many Arabic morphological analysers and generators have been developed in the last two decades particularly by the works of Tim Buckwalter [14,15], Habash and Rambow [16], the Linguistics Data Consortium [17], Sawalha [18] and Attia [12]. Recent works have attempted either to improve the accuracy of the analyser [19], to add some other features [20] or to focus on a specific dialect [21]. We argue that there is a gap between morphology and semantics in the current approaches to Arabic morphology and the focus has been placed only on the issues of word formation and decomposition. We cannot go higher than morphology. Morphologically speaking, the finite state approach [11] can be seen as relatively close to our approach. However, a finite state machine does not know units, it is just a sequence of states. Our approach has both a context free grammar on the one hand and a mechanics to deal with discontinuity on the other hand. This allows the computer to understand ⟨k, t, b⟩ and ⟨a, a⟩ as units, not states, because they have meanings. A clearer example can be seen, for example, in German where the "trennbare Präfix" (separable prefix), like /auf/ in /aufmachen/, and the verb appear in some cases as two parts. This can occur, for instance, in the imperative, as in "Mach bitte das Fenster auf!" (*Open the window, please.*). However, /mach/ and /auf/ together form one unit (as they can only be interpreted together) but this unit is discontinuous. This also applies to Arabic roots. Our approach is different from previous and current approaches in (i) it is compositional; morphological units have meanings and the meaning of complex expressions is determined by the meaning of these smaller units and their order; (ii) morphology is not distinguished from syntax. Rather, it is a lexicalist approach on a par with categorial grammar; (iii) it is more restrictive; since it purely surface-oriented, it allows only grammatical constituents. This eliminates the overgeneration.

3   A Purely Surface-Oriented Approach

At the simplest level of description, a natural language is simply a set of strings over a given alphabet A [22]. A^* denotes the set of all strings over A. The concatenation of two strings x and y is denoted by $x\,\hat{}\,y$ or simply xy. Concatenation is associative, that is, $(x\,\hat{}\,y)\,\hat{}\,z = x\,\hat{}\,(y\,\hat{}\,z)$, $\langle A^*, \hat{}, \varepsilon \rangle$ constitutes a monoid [22]. In the sequel, variables for strings are formed using a vector arrow, e.g. $\overrightarrow{x}$.

3.1   Glued Strings

We propose the notion "glued string" rather than just string. A glued string is a string with two context conditions: one for the left context and one for the right context. These conditions specify the properties of the string $\overrightarrow{x}$ such that $\overrightarrow{x}$ can appear in $\overrightarrow{u}\,\overrightarrow{x}\,\overrightarrow{v}$. A very good and clear example in Arabic is /ة/ ['p'], which is followed only by space (forget now about case markers) and consequently no other strings can follow it directly. This is a left-hand condition for the suffix and a right-hand condition for the /ة/. Moreover, the /ة/ is not allowed to be preceded by a space. So, before we can formally define glued strings we should, firstly, define the notion of a requirement. A **requirement** is a pair $(s, \overrightarrow{x})$, where s is a sign[3] and $\overrightarrow{x}$ a string. Roughly, the context must contain one of the strings with sign $+$ as suffix (if on the left) and as prefix (if on the right), while avoiding all the strings with sign $-$. We can now define the glued string as follows.

Definition 1 (Glued string). *A **glued string** is a triple $j = \langle L, \overrightarrow{x}, R \rangle$, where L is a set of left requirements, $\overrightarrow{x}$ is a string, and R is the set of right requirements.*

We give three explanatory examples of glued strings in English, Hungarian (taken from [3]) and in Arabic (/␣/ denotes the blank):

1. $j = \langle \{(+, \text{ch}), (+, \text{s}), (+, \text{sh}), (+, \text{x}), (+, \text{z})\}, \text{es}, \{(+, ␣)\} \rangle$
   This example codes the fact that the plural morph /es/ in English is suffixed only to words ending in ch, s, sh, x or z, while it must strictly be at the end of the word.

---

[3] There shouldn't be a confusion with the term sign as used in the abstract. A sign here is just plus $+$ or minus $-$.

2. $j_1 = \langle\{(+, b)\}, bal, \varnothing\rangle$

   This example from Hungarian specifies that the instrumental form /bal/ appears only after b.

3. The previous example/ ة/ with its right and left requirements as a glued string j_2: $j_2 = \langle\{(-, \smile)\}, ة, \{(+, \smile)\}\rangle$

3.2   Occurrence

A string occurs in another string if the first is a substring of the second. Let $\vec{x}$ and $\vec{y}$ be two strings. An occurrence of $\vec{x}$ in $\vec{y}$ is a pair $o = \langle\vec{u}, \vec{v}\rangle$ such that $\vec{y} = \vec{u}\,\vec{x}\,\vec{v}$. A similar definition of occurrence can be also found in [22]. For example, let $\vec{y} = /$ معلمون $/$ ['muEalolimUna', 'teachers'], $\vec{u} = /$ م $/$ ['mu'] and $\vec{v} = /$ ون $/$ ['Una']. Then $\vec{x} = $ علم ['Ealolim'].

If $o_1 = \langle\vec{u_1}, \vec{v_1}\rangle$ is an occurrence of $\vec{x_1}$ in $\vec{y}$ and $o_2 = \langle\vec{u_2}, \vec{v_2}\rangle$ is an occurrence of $\vec{x_2}$ in $\vec{y}$ then it is said that o_1 is to **the left of** of o_2 and (o_2 is to the **the right of** o_1) if $\vec{u_1}\vec{x_1}$ is a prefix of $\vec{u_2}$. If $\vec{u_2} = \vec{u_1}\vec{x_1}$, it is said that o_1 is **immediately to the left of** (left adjacent to) o_2 and o_2 is **immediately to the right of** (right adjacent to) o_1. The two occurrences o_1 and o_1 are said to be **contiguous** if one of them is **left** or **right adjacent** to the other. Otherwise, they **overlap**.

3.3   Morphological Class

The notion of a glued string has been developed to handle the conditions of combination. However, in many cases this notion doesn't suffice to handle all possible conditions. The feminine suffix in Arabic / ة / ['p'] is used in most cases to differentiate between masculine and feminine as in / معلم / ['muEalolim', 'teacher (male)'] and / معلمة / ['muEalolimap', 'teacher (female)'] respectively. However, in some cases this suffix is ignored because the noun or the adjective refers by nature to a female as in / طالق / ['TAliq', 'divorced'] and / حائض / ['HA'iD', 'menstruant']. Also, in some other cases the / ة / ['p'] is ignored because the same word is used to refer to persons of both genders as in / جريح / ['jarYH', 'injured'] and / قتيل / ['qatYl', 'killed'].

This means that the combination process depends in some cases on information that cannot be captured only by phonology. Therefore, we need an additional mechanism to capture such cases. This is the notion of "morphological class". These are properties of individual morphs, not morphemes, that control the behavior of a morph under combination. When two morphs are to be combined, one of them takes the role of the argument, while the second takes the role of the functor. When a morph m_1 takes a morph m_2 as its argument, the two give a third morph m_3. This can be written in a form of a function:

$$m_1(m_2) = m_3$$

In the combination process between root morph and pattern morph, for example, the pattern morph takes the role of the functor, let say as above m_1 and the root morph, m_2, takes the role of the argument. Their combination produces a stem or m_3 as in the following example:

$$m_1(m_2) = m_3$$

$$pattern(root) = stem$$

$$\langle a, a \rangle(\langle k, t, b \rangle) = /\text{katab}/$$

As we can see from the example, to properly handle the combination between roots and patterns, we have to first deal with the properties of m_1 and m_2 before the combination; i.e. to be combined. Then, we have the properties of m_1 and m_2 after combination; i.e. m_3. The combination in the previous example results in a perfective stem in the active form. Compare that, now, with the following combination:

$$m_1(m_2) = m_3$$

$$pattern(root) = stem$$

$$\langle u, i \rangle(\langle k, t, b \rangle) = /\text{kutib}/$$

The result of the combination, in this case, is the perfective stem in the passive form. This suggests that each morph that takes the role of the functor (m_1, or the pattern morph in these examples) has two classes, an "ingoing class" and the "outgoing class". The ingoing class specifies what class the argument (m_2, or the root morph) must have to combine with m_1. The outgoing class states what class the combination of m_1 and m_2 has. This mechanism has proven sufficient not only in allowing proper combination but also in preventing improper ones. The number of features that guide the combination can be large and the morphological classes themselves can be rather complex. Therefore, an attribute value matrix (AVM) in the following form is suggested:

$$\begin{bmatrix} ATTRIBUTE_1 : valueset_1 \\ ATTRIBUTE_2 : valueset_2 \\ \cdots \qquad \cdots \\ ATTRIBUTE_n : valueset_n \end{bmatrix}$$

If $n = 0$, the AVM is empty, and if $n \geq 1$, the $ATTRIBUTE_i$ are names of features such as case, gender, number etc. and the $valueset_i$ are sets of admissible values for each attribute. Here, given an attribute a, $rg(a)$ denotes the set of admissible values for a. So, we require $valueset_i \subseteq rg(ATTRIBUTE_i)$ for every $i \leq n$. For example, in Arabic $rg(NUM) = \{singular, dual, plural\}$. Hence, the following is a legal AVM.

$$\begin{bmatrix} NUM : \{dual, pl\} \end{bmatrix}$$

Sets of values encode underspecification. Using logical notation, we may write instead

$$\begin{bmatrix} NUM : dual \vee pl \end{bmatrix}$$

$\top$ denotes the set of all values. So we have

$$[\,NUM : \top\,] \equiv [\,NUM : \{sing, dual, pl\}\,]$$

where $\equiv$ denotes logical equivalence. Arabic nouns/adjectives, for instance, are inflected for case, state, gender and number. These four features have the following ranges:

$$rg(CASE) = \{nom, acc, gen\}$$
$$rg(STATE) = \{def, indef\}$$
$$rg(GEN) = \{masc, fem\}$$
$$rg(NUM) = \{sing, dual, pl\}$$

Conjunction and disjunction may be used to combine AVMs. The following equivalence is evident from the definition of AVMs.

$$\begin{bmatrix} NUM : & \{sing\} \\ CASE : & \{nom\} \end{bmatrix} \equiv [\,NUM : \{sing\}\,] \wedge [\,CASE : \{nom\}\,]$$

When an attribute receives the empty set as value, this means that we have an empty disjunction, which is defined to be false ($\bot$):

$$[\,CASE : \varnothing\,] \equiv [\,CASE : \bot\,]$$

We can apply the usual laws of logic. Consider, for example, two attributes (say, CASE and NUM) and use the law of distribution:

$$\begin{bmatrix} NUM : & sing \vee pl \\ CASE : & nom \end{bmatrix} \equiv \begin{bmatrix} NUM : & sing \\ CASE : & nom \end{bmatrix} \vee \begin{bmatrix} NUM : & pl \\ CASE : & nom \end{bmatrix} \equiv$$

$$[\,CASE : nom\,] \wedge ([\,NUM : sing\,] \vee [\,NUM : pl\,]) \equiv$$

$$([\,CASE : nom\,] \wedge [\,NUM : sing\,]) \vee ([\,CASE : nom\,] \wedge [\,NUM : pl\,])$$

Definition 2 (Feature space). *A **feature space** is a triple $\sigma = \langle A, V, rg \rangle$ such that A is a finite set of attributes, V is a finite set of values and $rg : A \to \wp(V)$ a function such that for all $a \in A$, $rg(a) \neq \varnothing$. For Arabic, as given above, we may put $A := \{CASE, NUM, GEN\}$, $V := \{nom, acc, gen, sing, dual, pl, masc, fem\}$.*

Abstractly, an AVM is a partial function f from attributes to sets of admissible values. If f is undefined on a, we may extend f by putting $f(a) := rg(a)$. Thus, f may also be considered a total function. Consider, for example, the AVM of nouns/adjectives that follow prepositions in Arabic (in the genitive case).

The prepositions behave as a functor that takes nouns/adjectives as argument regardless the number, gender or state and give nouns/adjectives in genitive case. Therefore, in this case, the $f(NUM)$, $f(GEN)$ and $f(STATE)$ are equal to the $rg(NUM)$, $rg(GEN)$ and $rg(STATE)$ as shown in the following AVM:

$$\begin{bmatrix} POS: & noun \vee adj \\ NUM: & \top \\ GEN: & \top \\ STATE: & \top \end{bmatrix}$$

So, by convention we may extend f to $f(STATE) = \{def, indef\}$.

3.4   Discontinuity, Reduplication and Handlers

Discontinuity is used in grammatical analysis to refer to the splitting of a construction by the insertion of another grammatical unit [23]. The concept of discontinuity is central for handling Arabic in a completely surface-oriented compositional approach because morphs are the meaningful units of speech but clearly are not continuous, contrary to what is the case in most languages. The plural is formed in English simply by concatenation e.g. of /dog/ and /s/ to get /dogs/. However, the process of Arabic word formation is rather complex; it is not just a sequential concatenation of morphs by placing them next to each other: the constituents can be discontinuous. Vowels and more consonants are inserted between, before and after the root consonants. The idea is clear when we consider, again, the triliteral (3 consonant) root /ك ت ب/ ['k t b'] and some of its derivatives like /كتب/ ['kataba', 'he wrote' or 'kutiba', 'was written'], /كاتب/ ['kAtib', 'writer' or 'kAtab', 'correspond with'] and /مكتوب/ ['makotUb', 'is written']. Both root morphs and pattern morphs are instances of **fractured glued strings**.

Definition 3 (Fractured glued string). *A **fractured glued string** is a sequence of glued strings. If $\gamma_0, \gamma_1, \cdots, \gamma_{m-1}$ are glued strings, then $g := \gamma_0 \otimes \gamma_1 \otimes \cdots \otimes \gamma_{m-1}$ denotes the fractured glued string, formed from the γ_i in this order. γ_i is called the ith section of g. m is called the dimension of g, referred to as $dim(g)$. The unique fractured string with dimension 0 is denoted by ζ.*

We can write the Arabic root as $k \otimes t \otimes b$ and the morph of the third person singular in the active form of the past tense $a \otimes a$. The content of a string is defined as:

Definition 4 (String content). *If $\gamma = \langle L, \overrightarrow{x}, R \rangle$ is a glued string, then $c(\gamma) = \overrightarrow{x}$. Furthermore, $c(\otimes_{i<n}\gamma_i) = c(\gamma_0)\hat{\ }c(\gamma_1)\hat{\ }...\hat{\ }c(\gamma_{n-1})$.*

Context free grammars are not equipped to describe discontinuity. To deal with discontinuous constituents or, more particularly, to combine two fractured glued strings, [3], following [24] suggests using a combinatorial function called **handler**. A handler can be defined as follows:

Definition 5 (Handler). A **handler** is a sequence H of sequences of pairs (i, b), where i is a natural number and b a boolean. The members of H are called its **sections**. A pair (i, b) is said to **occur** in H, in symbols $(i, b) \in H$, if there is a section of which (i, b) is some member. The pairs occurring in H are called its parts. Parts may have several occurrences. The result of applying H to two fractured strings g and h such that $g = \gamma_0 \otimes \gamma_1 \otimes \cdots \otimes \gamma_{m-1}$ and $h = \eta_0 \otimes \eta_1 \otimes \cdots \otimes \eta_{n-1}$ is defined as follows. Put:

$$(i, b)(g, h) = \begin{cases} \gamma_i & if\, b = true \\ \eta_i & else \end{cases}$$

Now, for the sequence $h_i = (i_0, b_0), (i_1, b_1), \cdots, (i_{p-1}, b_{p-1})$, we put

$$h_i(g, h) := (i_0, b_0)(g, h)\hat{\,}(i_1, b_1)(g, h)\hat{\,}\cdots\hat{\,}(i_{p-1}, b_{p-1})(g, h)$$

Finally, let $H = (h_0, h_1, \cdots h_{q-1})$ have q sections, then:

$$H(g, h) := h_0(g, h) \otimes h_1(g, h) \otimes \cdots \otimes h_{q-1}(g, h)$$

A handler is used if and only if it is proper. A proper handler is defined as:

Definition 6 (Proper Handler). A handler H is proper if for all numbers i, j and Booleans b, if H contains (i, b) and $j < i$ then H also contains (j, b). The dimension of a handler H is defined by:

$$dimH = (\{i : (i, true) \in H\}, \{i : (i, false) \in H\})$$

If H is proper, $dimH$ is a pair of numbers, such that:

$$0 \text{ is the empty set } \phi \text{ and } n + 1 = \{0, 1, ..., n\}.$$

A handler $H(g, h)$ is defined if and only if H is proper and $dimH = (dim(g), dim(h))$ i.e. if all sections of the two fractured strings are used in H. This combinatorial function allows a sequence of glued strings to be combined in diverse ways: forward concatenation, backward concatenation, forward wrapping, reduction, forward transfixation and, beyond the MCFGs [24], reduplication [25] as shown in the following examples:

- Forward Concatenation:
  Put $F := \langle\langle(0, true), (0, false)\rangle\rangle$.
  Then $F(\overrightarrow{x}, \overrightarrow{y}) = \overrightarrow{x}\,\overrightarrow{y}$
- Backward Concatenation:
  Put $B := \langle\langle(0, false), (0, true)\rangle\rangle$.
  Then $B(\overrightarrow{x}, \overrightarrow{y}) = \overrightarrow{y}\,\overrightarrow{x}$
- Forward Wrapping:
  Put $W := \langle\langle(0, true), (0, false), (1, true)\rangle\rangle$.
  Then $W(\overrightarrow{x} \otimes \overrightarrow{v}, \overrightarrow{y}) = \overrightarrow{x}\,\overrightarrow{y}\,\overrightarrow{v}$
- Reduction:
  Put $R := \langle\langle(0, true), (1, true)\rangle\rangle$.
  Then $R(\overrightarrow{x_0} \otimes \overrightarrow{x_1}) = \overrightarrow{x_0 x_1}$

– Transfixation:

Put $T := \langle\langle(0, true), (0, false), (1, true), (1, false)\rangle\rangle$.

Then $T\left(\overrightarrow{x_0} \otimes \overrightarrow{x_1}, \overrightarrow{y_0} \otimes \overrightarrow{y_1}\right) = \overrightarrow{x_0}\overrightarrow{y_0}\overrightarrow{x_1}\overrightarrow{y_1}$

We can see how this works with a sequence of glued strings in Arabic. If we want, for example, to form the word /كاتب/ ['kAtib', 'writer'] from the root $k \otimes t \otimes b$ and the vowels $A \otimes i$ we apply the following handler (note that the pattern morph plays the role of the functor). Put

$$H := \langle\langle(0, false), (0, true), (1, false), (1, true), (2, false)\rangle\rangle$$

If we apply a function that maps from each part of the handler to its corresponding string in the two fractured strings, we get the following:

– $(0, false)$ /k/   – $(1, false)$ /t/   – $(2, false)$ /b/
– $(0, true)$ /A/   – $(1, true)$ /i/

The result of applying this handler to the two fractured glued strings $k \otimes t \otimes b$ and $A \otimes i$ is /kAtib/ as shown:

$$H(A \otimes i, k \otimes t \otimes b) := k\hat{\ }A\hat{\ }t\hat{\ }i\hat{\ }b = kAtib$$

Reduplication is also an important feature in Arabic word formation. In some cases, Arabic tends to duplicate a specific letter (string) to get a new word as in /كتّب/ ['katotab', 'made sb write'] from /ك ت ب/ ['k t b'] or /درّس/ ['daroras', 'taught or educated'] from /د ر س/ ['d r s']. In this case, for example, reduplication results in not only a different form of the verb, but also a different meaning. Both /katab/ and /katotab/ have the same root /k t b/ and the same inserted vowels. However, the only difference between the two forms lies in the reduplication of the second consonant in the case of /katotab/ which changes the meaning from 'he wrote' to 'he made [sb] write'. This applies also to /daras/ and /daroras/. In some other cases, reduplication occurs in the root itself when the second consonant is doubled which means that the second and third consonants are the same as in /شدّ/ ['$dod'] and /مرّ/ ['mror']. Reduplication is represented orthographically with shaddah or tashdid above the duplicated letter (/شّ/ and /رّ/) and represented in Buckwalter's transliteration model as /~/ but we use instead two letters separated by sukoon because the duplicated letter is actually two letters; the first is followed by sukoon and the second by a vowel. This type is referred to as doubled or geminate verbs. Moreover, some Arabic four consonant roots are composed by the reduplication of the first two consonants twice as in /زلزل/ [z l z l] and /زعزع/ [z E z E].

Reduplication is not a unique feature of Arabic and it is also not the sole morphological operation. However, there are languages in which processes like reduplication is the primary morphological operation [26]. The plural formation in Malay is obtained, for instance, by doubling the singular form as in /orang/ "man" which becomes /orang-orang/ in the plural. The two parts are separated by a hyphen in the written language.

In multiple context free grammars [27], no component is allowed to appear in the value of the function more than once. This is not the case in the above-mentioned examples in Arabic and Malay. To handle such cases, the following handler for the plural form in Malay must be used.

$$D := \langle\langle(0, false), (0, true), (0, false)\rangle\rangle.$$

Then

$$D(\text{-}, \text{orang}) = \text{orang}\hat{}\text{-}\hat{}\text{orang}$$

Arabic is not an exception; we allow the handler to combine the multi occurrence of any substring but we insert a sukoon between the duplicated letter. Put

$$D := \langle\langle(0, false), (0, true), (1, false), (1, true), (1, false), (2, true)(2, false)\rangle\rangle$$

Then

$$D(a \otimes o \otimes a, k \otimes t \otimes b) = k\hat{}a\hat{}t\hat{}o\hat{}t\hat{}a\hat{}b = \text{katotab}$$

as well as

$$D(a \otimes o \otimes a, d \otimes r \otimes s) = d\hat{}a\hat{}r\hat{}o\hat{}r\hat{}a\hat{}s = \text{daroras}$$

However, in the real implementation, we deal with the reduplication in Arabic from a different perspective for the sake of simplification. We utilize the morphological class technique to sub-categorize the root into several types and each category is allowed to merge with specific patterns as discussed in details in the introduction.

3.5   Morphs and Morphemes

The actual units of expression of language are the morphs. They comprise three components. The first is the exponent, which is a fractured glued string. The second is a sequence of selectors, which determine what arguments the morph takes. And the third is a rank function. This function is only needed for empty morphs, to prohibit infinite derivations, and will concern us no further. Morphemes, which are the only meaning bearing units, are sets of morphs that share a common semantics. Thus, the lexical units pair meaning representations with morphemes, not morphs. This accounts for the fact that morphemes can have many different surface forms.

Definition 7 (selector). *A **selector** is defined formally as triple $\sigma = (M, N, H)$, where M and N are morphological classes and H is a handler. M is called the **in-class** of σ and N is its **out-class**.*

The role of the selector is to specify what happens when a morph is applied to another morph. The application of one morph (the functor) to another (the argument) is only defined if the in-class of the functor unifies with the out-class of the argument. Given, for example, a functor $\sigma = (M, N, H)$ and an argument $\sigma' = (M', N', H')$, the application of σ to σ' is defined in first instance as follows:

$$\sigma \cdot \sigma' := (M', N, H \circ H')$$

However, underspecification must be handled properly. The way this is standardly done is that the out-class is not actually underspecified, but is a function of its in-class, which is genuinely underspecified. Underspecified values in the out-class are copies of the actual in-class values. Indeed, the proper way to view selectors is as pairs (f, H) where f is a function from fully specified morphological classes to fully specified morphological classes, and H is a handler. This function is given by the pair of AVSs as follows. For each attribute ATT, the associated function f is the following.

– ATT is given a value a in M and a value b in N. Then f is defined on all nonempty values $a' \subseteq a$ and returns b.
– ATT is given no value in M but value b in N. Then f is defined on all nonempty values $a' \subseteq rg(\text{ATT})$ and returns b.
– ATT is given a value a in M but no value in N. Then f is defined on all nonempty values $a' \subseteq a$ and returns a'.
– ATT is neither given a value in M nor in N. Then f is defined on all $a \subseteq rg(\text{ATT})$ and returns a.

The product of (M, N) and (M', N') is specified by computing the values of each occurring attribute.

Definition 8 (Morpheme). *A **morpheme** is a set of morphs that share the same semantics but differ in the form.*

Arabic broken plural is, from this perspective, highly allomorphic; for a given singular pattern two different plural forms may be equally frequent, and sometimes, for some singulars as many as three further statistically minor patterns are also possible [28]. Given morphemes M an N, the combination $M \star N$ is the set of all $m(n)$ such that $m \in M$ and $n \in N$.

4 Arabic Morphology Within the Proposed Framework

In order to deal with the complexity of Arabic in a completely surface-oriented approach, we have to deal with two basic problems: root and pattern matching and morphophonemic alternations.

4.1 Root and Pattern Matching

Roots behave differently when they combine with patterns to form stems. Each root chooses specific vowels from the following six possibilities to form perfective and imperfective stems:

- (a, a) and (o, u)   - (a, a) and (o, a)   - (a, u) and (o, u)
- (a, a) and (o, i)   - (a, i) and (o, a)   - (a, i) and (o, i)

Arabic traditional grammarians attempted to handle this problem by classi-
fying roots into basic and sub-categories. That is, roots are classified depending
on the number of consonants into triliteral and quadrilateral. This subdivision
is not necessary since handlers need a fixed length. However, it is better to use
this as an explicit feature. Each category is further divided into different terms.
Glued strings are not sufficient because the phonology does not provide us with
enough information to determine which root interlock with which pattern. We
take advantage of the traditional categorization and sub-categorization of morphemes and utilize the notion of morphological classes. Particularly, each root
morph has a morphological class with four attributes:

- **Type**: This attribute differentiates between the two basic types of roots;
  triliteral and quadrilateral. This means that the **type** attribute in the morphological class of any root has a set of values that has two elements and
  consequently has the following range:

$$rg(TYPE) = \{tri, quad\}$$

- **Subtype**: This attribute divides Arabic roots depending on the nature of
  the letters of the root, not on their number, and position of specific letters
  within the root. Roots are divided into sound (free of semi-vowels, hamza
  and reduplication), geminate (with a duplicated letter), first hamzated (the
  first letter is hamza), second hamzated (the second letter is hamza), third
  hamzated (the third letter is hamza), assimilated (the first letter is either
  wAw or yA), hollow (the second letter is either wAw or yA), defective (the
  third letter is either wAw or yA), first weak (the first and third letter are
  either wAw or yA) and second weak (the second and third letter are either
  wAw or yA). Thus, the **Subtype** attribute has a range of ten elements.
- **Form**: As mentioned before, Arabic roots, in terms of the numbers of letters,
  are divided into triliteral and quadrilateral. However, other consonants can be
  added to the two basic types. The triliteral roots can host up to three other
  consonants while the quadrilateral roots can have only two more consonants.
  In practice, not every lexical root occurs in all different forms but they vary
  from one another and dictionaries normally list all the forms in which a
  lexical root regularly appears [4]. If the root consists of only the three or four
  consonants, it is in the base form called form I, which is also referred to in
  Arabic as mujarorad, literally the "stripped" form; otherwise it is in one of
  the forms II to X, which are referred to as mazId, literally, "increased" forms,
  i.e., more morphologically complex [4]. Therefore, in terms of the form, the
  root takes a Roman number extending form I to X in case of triliteral roots
  and from I to IV in case of quadrilateral roots. Thus the **form** attribute in
  the morphological class has the following range:

$$rg(FORM) = \{I, II, III, \cdots, X\}$$

– **Stem-Eayon:** It is traditional to refer to the short vowel which follows the second root consonant of a verb as the "stem vowel" [4] and we saw in the beginning of this section that the stem vowel (Eayon) may vary from one stem to another. We capture these possibilities using an attribute with a value set that has six values. These values specify the stem vowel in perfective and imperfective forms and therefore this attribute has the following range: $rg(Stem - Eayon) = \{au, ai, aa, ia, uu, ii\}$.

The morphological class of the root $\langle k, t, b \rangle$

$$\begin{bmatrix} TYPE: & tri \\ SUBTYPE: & sound \\ FORM: & I \\ STEM\text{-}EAYON: & au \end{bmatrix}$$

On the other hand, pattern morphs are provided with an in-class (for the hosted root morph) and an out-class (for the result of the merge). The merge is defined only if the out-class of the root matches the in-class of the pattern. Consider, for example, the in-class of the pattern morph $\langle o, u \rangle$.

$$\begin{bmatrix} TYPE: & tri \\ SUBTYPE: & \{sound, shamzated, shamzated, thamzated\} \\ FORM: & I \\ STEM\text{-}EAYON: & au \end{bmatrix}$$

The out-class of the root morph and in-class of the pattern morph match and can be merged using this handler:

$$H := \langle\langle(0, false), (0, true), (1, false), (1, true), (2, false)\rangle\rangle$$

$$H (o \otimes u, k \otimes t \otimes b) := k\hat{}o\hat{}t\hat{}u\hat{}b = kotub$$

This applies to the pattern morph o⊗i which merges with the root morph d⊗r⊗b to give the imperfective stem /dorib/. It is now clear that both */dorub/ and */kotib/ are not allowed because neither the pattern morph $o \otimes u$ can merge with the root morph $d \otimes r \otimes b$ nor the pattern morph $o \otimes i$ is allowed to merge with the root morph $k \otimes t \otimes b$. This is because the associated classes, in either cases, do not match.

4.2   Morphophonemic Alternations

The interaction among different type of morphs may result in phonological alternations. Arabic roots are broadly classified into two types depending on the presence or the absence of the wAw and yA': sound and weak. Weak roots, in particular, may undergo stem changes when inflected. Let's consider, for example, the following derived forms of the same hollow root / ع ي ب / $b \otimes y \otimes E$:

- /باع/ ['bAEa', 'he sold'], masculine third person singular in the perfective form.
- /بعت/ ['biEotu', 'I sold'], masculine or feminine first person singular in the perfective form.
- /يبيع/ ['yabYEu', 'He sells'], masculine third person singular in the imperfective form.

Arabic grammarians attribute such morphophonemic changes to the rule of "origin of Alif" in forms like /باع/ ['bAEa'] from /ب ي ع/ $b \otimes y \otimes E$ and /قال/ ['qAla'] from /ق و ل/ $q \otimes w \otimes l$. They argue that the "alif" is returned to its original "yA'" or "wAw", however, this says nothing about the dropping of the second letter of the root in /بعت/ ['biEotu'] for instance. Anyway, in our approach, we are not concerned about the rules that justify such changes because in order to ensure the purely surface-oriented treatment of Arabic word formation, no rule is allowed to delete or modify any string. That is, we cannot say, for example, that the alif of the hollow verb is changed to its origin yA' or wAw in specific forms. Therefore, we don't add a rule to modify the alif of /قال/ ['qAla'] to its original wAw and another rule to drop it in /قلت/ ['qulotu'] and this applies also to many cases in which the interaction among different morphs may result in some changes. Instead, each root is a morpheme, that is, a set of morphs that correspond to all possible forms under merge. Thus, for the root morpheme /ب ي ع/ $b \otimes y \otimes E$ we may have up to three morphs:

- /ب ي ع/ $b \otimes Y \otimes E$, from which we can get /يبيع/ ['yabYEu', 'he sells'], /بيعوا/ ['bYEUA', 'sell'] etc.
- /ب ع/ $b \otimes E$, from which we can get /بعت/ ['biEotu', 'I sold'], /يبع/ ['yabiEo', 'sell'] etc.
- /ب ا ع/ $b \otimes A \otimes E$, from which we can get /باع/ ['bAEa', 'he sold'], /باعوا/ ['bAEUA', 'they sold'] etc.

This permits not only to get the grammatical forms but also to disallow the ungrammatical ones like * /ياعوا/ ['yabAEu']. So far we dealt with challenges that Arabic morphology poses to this approach. We have already developed a lexicon for a fragment of Arabic. It is true that the lexicon is small and there are many other cases of morphophonemic changes, however, they can be handled in the same way.

5   Results and Future Work

In this paper, we presented a mathematical complete surface-oriented approach to handle Arabic morphology. In this approach, we dealt with the language as a sequence of strings that are only allowed to be concatenated and reduplicated but no rule is allowed to add, remove or modify any string. At the very beginning, we reviewed briefly different theoretical and computational approaches to

Arabic. We argue that there is a gap in current approaches between morphology and semantics. In order to handle Arabic morphology within this approach, we dealt with two problems: root and pattern matching and morphophonemic alternations. We have already developed a lexicon for a fragment of Arabic, and in the present time we extend our work on Arabic morphology in connection with the demands set by compositional semantics planning to come up with a morphosemantic lexicon for a fragment of Arabic.

References

1. CIA: CIA World Fact Book. Central Intelligence Agency, Washington, D.C. (2018)
2. Habash, N.: Introduction to Arabic natural language processing. Morgan and Claypool Publishers (2010)
3. Kracht, M.: Agreement morphology, argument structure and syntax. Revision 8 (2016, unpublished manuscript)
4. Ryding, K.: A Reference Grammar of Modern Standard Arabic. Cambridge University Press, Cambridge (2005)
5. Al-Sughaiyer, I., Al-Kharashi, I.: Arabic morphological analysis techniques: a comprehensive survey. J. Assoc. Inf. Sci. Technol. **55**(3), 189–213 (2004)
6. Soudi, A., Neumann, G., Van den Bosch, A.: Arabic Computational Morphology: Knowledge-Based and Empirical Methods. Springer, Cham (2007). https://doi.org/10.1007/978-1-4020-6046-5
7. Dichy, J., Farghaly, A.: Grammar-lexis relations in the computational morphology of Arabic. In: Soudi, A., Neumann, G., Van den Bosch, A. (eds.) Arabic Computational Morphology: Knowledge-based and Empirical Methods, pp. 115–140. Springer, Dordrecht (2007). https://doi.org/10.1007/978-1-4020-6046-5_7
8. Boudchiche, M., et al.: AlKhalil Morpho Sys 2: a robust Arabic morpho-syntactic analyzer. J. King Saud Univ.-Comput. Inf. Sci. **29**(2), 141–146 (2017)
9. Sawalha, M., Atwell, E.: Comparative evaluation of Arabic language morphological analysers and stemmers. In: Coling 2008: Companion volume: Posters, pp. 107–110 (2008)
10. Kay, M.: Nonconcatenative finite-state morphology. In: Proceedings of the Third Conference of the European chapter of the Association for Computational Linguistics, pp. 2–10 (1987)
11. Beesley, K.: Finite-state morphological analysis and generation of arabic at xerox research: status and plans in 2001. In: ACL Workshop on Arabic Language Processing: Status and Perspective, pp. 1–8 (2001)
12. Attia, M., et al.: A corpus-based finite-state morphological toolkit for contemporary Arabic. J. Logic Comput. **24**(2), 455–472 (2013)
13. Aboamer, Y., Farghaly, A.: Mariam ComLex: A Bi-Directional Finite State Morphological Transducer for MSA. In: The 29th Annual Symposium on Arabic Linguistics, at the University of Wisconsin-Milwaukee, USA (2015)
14. Buckwalter, T.: Buckwalter Arabic Morphological Analyzer, Version 1.0. Linguistic Data Consortium, University of Pennsylvania, LDC Catalog No: LDC 2002 L49 (2002). ISBN 1-58563-257-0
15. Buckwalter, T.: Buckwalter Arabic Morphological Analyzer, Version 2.0. Linguistic Data Consortium, University of Pennsylvania, LDC Catalog No: LDC 2004 L02 (2004). ISBN 1-58563-324-0

16. Habash, N., Rambow, O, Roth, R.: MADA + TOKAN: a toolkit for arabic tokenization, diacritization, morphological disambiguation, POS tagging, stemming and lemmatization. In: Proceedings of the 2nd International Conference on Arabic Language Resources and Tools (MEDAR), Cairo, Egypt (2009)
17. Maamouri, M., et al.: LDC Standard Arabic morphological analyzer SAMA v. 3.1. Linguistic Data Consortium, University of Pennsylvania, LDC Catalog No. LDC2010L01. ISBN 1-58563-555-3
18. Sawalha, M., Atwell, E., Abushariah, M.: SALMA: standard arabic language morphological analysis. In: 1st International Conference on Communications, Signal Processing, and their Applications (ICCSPA), pp. 1–6 (2013)
19. Abdelali, A., et al.: Farasa: a fast and furious segmenter for Arabic. In: Proceedings of the 2016 Conference of the North American Chapter of the Association for Computational Linguistics: Demonstrations, pp. 11–16 (2016)
20. Taji, D., et al.: An Arabic morphological analyzer and generator with copious features. In: Proceedings of the Fifteenth Workshop on Computational Research in Phonetics, Phonology, and Morphology, pp. 140–150 (2018)
21. Habash, N., Eskander, R., Hawwari, A.: A morphological analyzer for Egyptian Arabic. In: Proceedings of the Twelfth Meeting of the Special Interest Group on Computational Morphology and Phonology, pp. 1–9 (2012)
22. Partee, B., ter Meulen, A., Wall, R.: Mathematical Methods in Linguistic. Linguistic Society of America (1990)
23. Crystal, D.: A Dictionary of Linguistics and Phonetics, 6th edn. Blackwell Publishing Ltd. (2008)
24. Seki, H., et al.: On multiple context-free grammars. Theor. Comput. Sci. 88(2), 191–229 (1991)
25. Kracht, M., Aboamer, Y.: Argument structure and referent systems. In: 12th International Conference on Computational Semantics IWCS (2017)
26. McCarthy, J.: A prosodic theory of nonconcatenative morphology. Linguist. Inquiry 12(3), 373–418 (1981)
27. Kasami, T., Seki, H., Fujii, M.: Generalized Context-free Grammars, Multiple Context-free Grammars and Head Grammars. Preprint of WG on Natural Language of IPSJ (1987)
28. Soudi, A., Violetta C., Jamari, A.: The Arabic noun system generation. In: Proceedings of the International Symposium on the Processing of Arabic (2002)

A Topos-Based Approach to Building Language Ontologies

William Babonnaud[✉]

LORIA, Université de Lorraine, CNRS, Inria Nancy Grand Est, Nancy, France
william.babonnaud@loria.fr

Abstract. A common tendency in lexical semantics is to assume the existence of a hierarchy of types for fine-grained analyses of semantic phenomena. This paper provides a formal account of the existence of such a structure. A type system based on the categorical notion of topos is introduced, and is shown to be possibly adaptable to several existing formal approaches where such hierarchies are used. A refinement of the type hierarchy based on Fred Sommers' ontological theory is also proposed.

Keywords: Formal semantics · Lexical semantics · Type theory · Type ontology · Category theory

1 Introduction

The work presented in this paper originates from a very general question: *what is a semantic type?* Although a definitive answer to this complex, almost philosophical question will not be provided here, it is still possible to look for a better understanding of the role of semantic types in formal and lexical semantics by browsing the history of their use in computational linguistics. If this field was to be described as a meeting of computer science and linguistics, unsurprisingly there would be various contributions to the notion of semantic type from each discipline; and indeed it will be argued that this notion seems to sum up contributions of different origins which happened to work well together.

The notion of type has been ubiquitous in logics and computer science since the early works of Russell, which paved the way for the simple type theory of Church [7]. This theory was then introduced in computational linguistics by the founding work of Montague [14], who merely used it as a formal tool to embed the Fregean view of predicates as functions into his grammar. Thus he provided a formal basis for the *principle of compositionality,* which states that the meaning of an expression can be derived from the meanings of its constituent parts and from the way these parts are syntactically combined. The type system used in this approach is a very minimalist one, directly inherited from Church's theory, with a type e for entities and a type t for propositions.[1]

---

[1] Which correspond respectively to ι and o in Church's notation. Although it is worth studying, the additional s type, denoting intension, will not be discussed in this paper, for it is assumed to be a necessary feature for the set-theoretic models of the logic used by Montague. The conception of types as denotations of sets will be overlooked here as it is too specific.

© Springer-Verlag GmbH Germany, part of Springer Nature 2019
R. Bernardi et al. (Eds.): FG 2019, LNCS 11668, pp. 18–34, 2019.
https://doi.org/10.1007/978-3-662-59648-7_2

Following Montague's idea, those types can be given a practical interpretation: t denotes everything that can have a truth value, and e denotes individuals, which could be understood as the more basic expressions which can be used as arguments to another expression. This approach has been widely adopted as the foundation of formal semantics, and even if Montague's categories have been separated between syntactic and semantic ones, the use of a type for truth values and at least another one for predicate arguments, as well as the correspondence between syntactic categories and semantic types, are well-established.

However, a relatively recent tendency in computational semantics is to postulate a whole hierarchy of types of arguments, instead of a single one for entities, in order to account for specific semantic phenomena in a more fine-grained way. Such a hierarchy is intended to encode hypernymic relations between nouns, that is, inclusions between the "sets" (in a non-mathematical acceptance of the word) of entities satisfying the properties described by those nouns. Constructing hypernymic relations is exactly what we do when we make generalising statements, for instance:

(1)   a.   A dog is a mammal.
        b.   A mammal is an animal.
        c.   An animal is a living entity.

The type hierarchy thus reflects a part of common (yet *a priori* language-dependent) lexical knowledge; and far from being an abstract construction, such a structure can be acquired on the basis of an extensive empirical analysis as is done in the case of building hypernymic and hyponymic relations of synsets in the WordNet database [13].[2] This structure is at the core of lexical semantics, as it is generally assumed whenever type ontologies are needed. The minimal use of such a hierarchy enables one to formally encode the semantic presuppositions of a given entity, or the ones corresponding to the argument of another predicate.

Prior to the systematisation of this idea, philosophers, psychologists, and linguists came up with the notion of *semantic category,* which can arguably be seen as the conceptual ancestor of the current understanding of the notion of type. This notion arose notably in the early work of Husserl, with certain influence on the Polish school through Leśniewski and Ajdukiewicz. Yet to the best of our knowledge, one of the first proposals for a hierarchical structure of such categories was sketched by Chomsky [6]. His categories organised expressions in a tree structure which aimed at allowing for degrees in grammaticality judgements. Kiefer [10] elaborated on this idea while separating grammatical and semantic categories, and ordered the latter by means of so-called *esse-relations,* that is, categorial inclusions as in (1). It is also worth noticing that a similar notion of semantic category is used in the field of psycholinguistics, starting with the early work of Wittgenstein [25] on the process of categorisation, which has led to relevant theories on language acquisition [4] and the concept of prototype [8,17], among others.

---

[2] See the website https://wordnet.princeton.edu/. The online application provides the opportunity to explore the hierarchy by browsing a word, selecting its synset and accessing the list of hyponyms and hypernyms.

Another major contribution comes from the philosophical work of Fred Sommers. For a dozen years he has been interested in language ontologies and elaborated a complete theory of ontology through a series of articles (see [20–22] among others for a partial, yet wide-covering compendium). He examined Russell's notion of type as well as Ryle's idea of *category mistake* to propose a theory based on the question of predication. The core of this theory lies in the notion of *spanning:* a predicate is said to span an entity when it is predicable truly or falsely to this entity, but not absurdly. For instance, the predicate 'philosopher' spans Socrates and Julius Caesar, but not the Eiffel tower, because it is absurd to wonder whether the Eiffel tower is a philosopher or not, as it cannot be decided. Thus, Sommers rejected the Quinean approach of considering every absurd sentence as logically false. Instead, he proposed an analysis of the correctness of a sentence on different levels: on the first level is the question of "grammaticality", that is, syntactic well-formedness; on the second level is "category correctness", i.e. whether the predicates in the sentence are applied in a non-absurd way, without category mistakes in the sense of Ryle; and then, the third level is the question of "consistence", i.e. whether the sentence avoid contradiction or not. The fourth level consists eventually in determining the empirical truth of the sentence w.r.t. the context. Each level of analysis can be treated only under the condition that the sentence was correct for the previous levels; which means that we can give a truth value to a propositional sentence only if it has no category mistake.

Sommers' idea was then to use the predicates to recover the Russellian notion of type: considering a predicate P in the language, he defines $|P|$ to be the class of things that are spanned by P. Such classes are referred to as ontological classes, or *categories.* It appears then that different predicates can define the same categories, as for instance 'sad' and 'angry'; he also defined types of predicates, or *A-types,* so that two predicates are of the same A-type if they define the same category. Ontology being "the science of categories", Sommers' definitions create a language ontology comparable to a skeleton of the language (seen as the collection of its predicates), by quotienting it by the relation of "defining the same categories." But at this stage the onlotogy lacks structure. Sommers remedied this by formulating a structural principle:

"If C_1 and C_2 are any two categories, then either C_1 and C_2 have no members in common or C_1 is included in C_2 or C_2 is included in C_1." [21, p. 355]

A lot of Sommers' works, notably in [20], aimed at establishing this law. It has many important consequences w.r.t. the ontology itself, as well as on some metaphysical questions. Inclusions of categories can be presented from the predicative point of view by defining the relation of *predicability:* P is predicable of Q if anything that is spanned by Q is also spanned by P. Then the isomorphism between predicability and category inclusion is quite clear. Sommers also noticed that the vocabulary of any language is finite, and so is the number of categories we can define in its ontology. Thanks to finiteness and the structural principle, Sommers eventually showed that there is a category which includes all categories, and that there are categories which do not include any other. Thus, he gives to his ontology the shape of a tree, which is similar to the current type hierarchies. To make sure that this structure is correct, he also provided a way to handle equivocal predicates, such as 'hard' in (2), by splitting their different meanings in appropriate places on the tree.

(2)   a.   The chair is hard.
        b.   The question is hard.

Yet having a tree structure is a stronger assumption than just being a lattice, and Suzman [23] pointed out the idea that Sommers' ontology fails to account for entities that could be seen as being in the intersection of categories which are not included one in the other. Actually, Sommers discussed this point in [22] and admitted that imposing a tree structure on his ontology makes some entities lie outside of it. Amongst those entities are the absurd ones, such as 'red numbers', but there is also another sort which is worth dwelling upon: those entities which Sommers called *heterotypical composites,* and which are built upon two or more elementary categories from the ontology. The word 'Italy' in sentence (3a) is an example of such a composite. Heterotypical composites differ from equivocal predicates in the fact that contrary to the latter, there is no zeugmatic effect when combining simultaneously the former with two predicates of different A-types, as illustrated by comparing sentences in (3) below:

(3)   a.   Italy is sunny and democratic.
        b.   #The chair and the question are hard.

Thus Sommers proposed a way for catching any expression in its ontology. The influence of his work on the current understanding of type hierarchies remains unclear, but in any case, he proposed a powerful and elegant ontological theory which is worth shedding light on.

This sketch of history helps to draw an overview of some influences on the notion of type: on one hand, formal semantics needs a type of truth values and functional types in order to compose the lexical units properly, and on the other hand lexical semantics needs a structured ontology of types to get a more refined analysis of semantic phenomena. The combination of both approaches leads us to a rather standard approach in computational semantics: a type system *à la* Montague where the type e has been replaced by more precise types from an ontological hierarchy. Recent works using this kind of approach include Luo's theory of coercive subtyping (e.g. [12]), Pustejovsky and Asher's type composition logic (TCL) [1,2] and also Retoré's ΛTy_n framework [16].

The remainder of this article will sketch the basis of a type system where a hierarchy of entity types is properly integrated into the usual compositional framework. This type system is intended to reflect the ontology of language, and rests upon a specific model from category theory, so that it should naturally lead to a λ-calculus. Although the complete calculus will not be defined here, a few lines of what properties can be expected from such a system will be given. Section 2 introduces necessary categorical tools for understanding topos theory;

then the type system and its properties will be described in Sect. 3. Finally, a short review of related works and concluding remarks will be given in Sect. 4.

2   A Synopsis of Topos Theory

Category theory can be presented as a formalisation of mathematical structures, and is known to have strong connections with typed λ-calculi (see e.g. [3,19]). In this regard, a type system can be given a categorical[3] semantic model. How this kind of correspondence works will not be explored here. Rather, this section aims to give some basic tools to understand the notion of *topos,* which refers to categories with specific properties that will be used extensively in the rest of this paper. Only necessary notions will be introduced here for questions of space; more details can be found e.g. in Robert Goldblatt's book [9].

Definition 1. *A category $\mathcal{C}$ is a class of objects $\mathcal{O}(\mathcal{C})$, and for every pair of objects $A, B \in \mathcal{O}(\mathcal{C})$ a class of morphisms (or arrows) $\mathcal{C}(A, B)$, and a composition operator on morphisms $\circ$ such that:*

- *for all $f \in \mathcal{C}(A, B)$ and $g \in \mathcal{C}(B, C)$, there is a morphism $g \circ f \in \mathcal{C}(A, C)$;*
- *the composition is associative, i.e. $h \circ (g \circ f) = (h \circ g) \circ f$;*
- *for all object A, there is an identity morphism $\mathrm{id}_A \in \mathcal{C}(A, A)$ which is neutral for composition, i.e. $g \circ \mathrm{id}_A = g$ and $\mathrm{id}_A \circ h = h$ for all relevant g and h.*

A well-known example of a category is Set, whose objects are sets and whose morphisms are functions between sets, equipped with the usual notion of composition. The notation $f \in \mathcal{C}(A, B)$ will be replaced by $f : A \to B$ whenever $\mathcal{C}$ in understood. Elaborating from this definition, categories can be "enriched" by defining objects with specific properties (usually gathered under the term *universal property*). The prototypical objects with such properties are initial and terminal objects, as given in the following definition.

Definition 2. *A terminal (resp. initial) object in a category $\mathcal{C}$ is an object 1 (resp. 0) such that for all $A \in \mathcal{O}(\mathcal{C})$, there is exactly one morphism $!_A : A \to 1$ (resp. $0_A : 0 \to A$).*

It is worth noticing that the objects introduced in this definition (and in subsequent ones as well) have the property to exist *up to isomorphism.* Two objects A and B are isomorphic (noted $A \cong B$) when there are morphisms $f : A \to B$ and $g : B \to A$ such that $g \circ f = \mathrm{id}_A$ and $f \circ g = \mathrm{id}_B$. Objects satisfying the universal property are generally not unique, but all of them can be proven to be isomorphic. Yet as this property is known it is common to refer to *the* terminal object instead of *a,* and similarly for other objects. In Set, any singleton is a terminal object, and the empty set is the initial object.

---

[3] As a convention, I will distinguish between the adjective *categorial* when talking about linguistics categories such as Chomsky's or Sommers', and the adjective *categorical* when talking about things from category theory.

Other relevant objects are products and pullbacks, which we define below. The second definition uses the notion of *diagram,* which can be understood as a representation of a small subpart of a category as a graph whose nodes are objects and edges are morphisms. A diagram is said to be *commutative* if all paths of morphism compositions between any two objects are equal.

Definition 3. *A* product *of two objects A and B is an object $A \times B$ equipped with two projections $\pi_1 : A \times B \to A$ and $\pi_2 : A \times B \to B$, such that for all objects C and morphisms $f : C \to A$ and $g : C \to B$ there is a unique morphism $\langle f, g \rangle : C \to A \times B$ satisfying $\pi_1 \circ \langle f, g \rangle = f$ and $\pi_2 \circ \langle f, g \rangle = g$.*

Definition 4. *A* pullback *of two morphisms of the same codomain $f : A \to D$ and $g : B \to D$ in an object $A \times_D B$ equipped with two morphisms $f' : A \times_D B \to B$ and $g' : A \times_D B \to A$, such that for all objects C and morphisms $h : C \to A$ and $k : C \to B$ satisfying $f \circ h = g \circ k$, there is a unique morphism $l : C \to A \times_D B$ such that the following diagram commutes:*

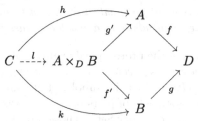

Another common kind of objects which appears in the definition of topoi are *exponentials,* but they will not be used in this paper, so mentioning their existence is enough. There is still a useful notion left, namely the *subobject classifier.* But first the notion of monomorphism should be defined:

Definition 5. *A morphism $f : A \to B$ is said to be a* monomorphism *(noted $f : A \rightarrowtail B$) if it is left-cancellable, i.e. if for any pair $g, g' : C \to A$ of arrows of codomain A, $f \circ g = f \circ g'$ implies $g = g'$.*

Monomorphism is actually a generalisation upon of notion of injective function, so that in Set monomorphisms are exactly those functions. Then, a *subobject* of an object A is an object B along with a monomorphism $B \rightarrowtail A$. Without loss of generality, it will be assumed that whenever B is a subobject of A, the associated monomorphism is unique. Then we can introduce properly the subobject classifier:

Definition 6. *A* subobject classifier *is an object Ω along with a monomorphism $\top : 1 \rightarrowtail \Omega$ such that, for all objects A and subobjects $B \rightarrowtail A$ of A, there is a unique morphism χ (called the* character *of B in A) making the following diagram a pullback (i.e. B is the pullback object of χ and $\top$):*

$$
\begin{array}{ccc}
B & \xrightarrow{\;!_A\;} & 1 \\
\Big\downarrow & & \Big\downarrow{\scriptstyle \top} \\
A & \xrightarrow{\;\chi\;} & \Omega
\end{array}
$$

The name *character* given to χ should help understand how the subobject classifier works: it represents an "object of truth values" and $\top$ corresponds to the value "true", so that any subobject can be associated with a morphism that distinguishes it by sending it on "true". In Set, any two-element set is a subobject classifier (again this notion is given up to isomorphism), and characters are exactly characteristic functions. Yet the previous definition imposes characters to be uniquely associated with subobjects, which is actually a very strong assumption. For any object A of a category $\mathcal{C}$ with subobject classifier, let call Sub(A) the class of subobjects of A.[4] The definition above imposes then the following property, which is usually referred to as Ω-*axiom:*

Proposition 1 (Ω-axiom). *For all $A \in \mathcal{O}(\mathcal{C})$, we have* Sub($A$) $\cong \mathcal{C}(A, \Omega)$.

We have then all tools in hand for introducing the central categorical notion on which is based this paper:

Definition 7. *A topos is a category with initial and terminal objects, all products, all pullbacks, all exponentials, and a subobject classifier.*

The category Set is again the prototypical example of a topos, which satisfies two additional properties: it is *bivalent*, i.e. it has exactly two distinct morphisms $1 \to \Omega$, and it is *classical*, i.e. its subobject classifier is isomorphic to the coproduct $1 + 1$. The category Set^2 of pairs of sets is an example of classical but non-bivalent topos, while the category $\mathsf{Set}^\to$ whose objects are functions between sets is neither classical nor bivalent. Also, if M is a monoid which is not a group, then the category M-Set whose objects are sets together with an action of M on them is an example of non-classical but bivalent topos.

There are many other equivalent ways for defining a topos, but the one given above rests exclusively on the notions that will be used in the remainder of this paper. As a subobject classifier is an object of truth values, there are ways in a topos for introducing an internal logic: we can define morphisms $\bot : 1 \to \Omega$, $\neg : \Omega \to \Omega$ and $\wedge, \vee, \Rightarrow$ as morphisms $\Omega \times \Omega \to \Omega$. This internal logic can serve as semantic model for intuitionistic logic, see [9] for details. Using such morphisms with the Ω-axiom allows to define new kind of subobjects. If B and C are subobjects of A of character χ_B and χ_C, then $\bar{B}$, $B \cup C$, $B \cap C$ and $B \Mapsto C$ are subobjects of A of respective characters $\neg \circ \chi_B$, $\vee \circ \langle \chi_B, \chi_C \rangle$, $\wedge \circ \langle \chi_B, \chi_C \rangle$ and $\Rightarrow \circ \langle \chi_B, \chi_C \rangle$. Those new operators have a particular behaviour in Sub(A):

Proposition 2. *In any topos, $\langle$Sub(A)$, 0, A, \cap, \cup, \Mapsto\rangle$ is a Heyting algebra for the subobject ordering.*

Also, the exponential object of A and Ω will be noted $\mathscr{P}(A)$. It is referred to as *powerobject* of A, and corresponds to powersets in Set. It has also interesting properties; some of them will be introduced in due time in the next section.

---

[4] For linguistically motivated reasons, classes of subobjects of an object and classes of morphisms between two objects will be considered as sets in the rest of this paper.

3 Building a Topos-Based Type System

This work is mainly inspired by the proposition of categorical model for TCL sketched by Asher in [1]. As explained before, there is a strong relation between typed λ-calculi and categories: we can use objects of the category for representing types, and morphisms for λ-terms. This approach is also motivated by the potential usefulness of categorical models to ensure the consistency of a compositional framework. In [1], his high need of pullbacks and powerobjects leads Asher to propose topoi as type models. Prior to his approach, topoi had been used for linguistic and cognitive questions, for instance in [11].[5] The choice of topos as the categorical basis of this approach is actually motivated by theoretical and practical concerns: the necessity of truth values in language semantics and of subtyping in a language ontology naturally suggests the use of a subobject classifier, which also requires a terminal object; and in order to enable the definition of a typed λ-calculus from such a category, it has to be at least cartesian closed,[6] that is, to have products and exponentials. Finally, as the aim is ultimately to build logical formulae for representing language semantics, we need an access to the usual quantifiers; and despite the subtleties that rule the distinction between the internal and external logics of the category, it is easier for now to add all pullbacks in it. The combination of these various requirements meets the Definition 7, so that topos is the minimal categorical structure needed for our purposes.

Building up on this idea, the theory developed here takes a closer look at how the properties of topoi can lead to a new type system for natural language semantics. More precisely, a specific instance of a topos will be introduced, and some properties it has to satisfy in order to define a type ontology for natural language will be described. It will also be argued that the type system thus created shares many properties with Sommers' propositions, which is an originally unexpected but welcome result.

3.1 From Montague's e to a Hierarchy of Types

Let $\mathcal{T}$ be a topos. Following the general idea of objects as types, we would like $\mathcal{T}$ to have at least the two Montagovian types e and t. The key to this model is the way we handle the type of truth values: indeed, by definition $\mathcal{T}$ has an object which corresponds exactly to what this type is supposed to be, namely the subobject classifier. Let therefore T be the subobject classifier of $\mathcal{T}$, and let E be a distinguished object of the topos. Following Montague [14], the first-order monadic properties (including nouns) are to be considered as terms of type E $\rightarrow$ T. Put in $\mathcal{T}$ those terms correspond to morphisms, so that each first-order monadic predicate in the language has a counterpart in $\mathcal{T}(\text{E}, \text{T})$, and conversely. Recalling the Ω-axiom, we get for each predicate a subobject of E. For instance,

---

[5] I am grateful to an anonymous reviewer for bringing this work to my attention.

[6] Actually, a monoidal closed category would suffice if we wanted a linear λ-calculus, but such a restriction is not justified here.

the predicate 'cat' defines the morphism **cat** : $E \to T$, and enforces by axiom the existence of a subobject CAT of E, so that the following diagram is a pullback:

As we assimilate objects of the topos and semantic types, the type CAT thus defined is exactly what we would expect: the type of entities that are actually cats, that is, those entities that are true of the predicate 'cat'. In our topos, this can be established by the equality of morphism compositions in the diagram above. For any object A of $\mathcal{T}$, define $\mathrm{true}_A : A \to T$ to be $\top \circ !_A$.[7] Then, the pullback above gives us the following equality:

(4) $\mathbf{cat} \circ f = \mathrm{true}_{\mathrm{CAT}}$

We can even do better if we introduce the notion of *global elements*. A global element of an object A in any category is a morphism $1 \to A$. Thus, the morphisms $\top$ and $\bot$ in our topos are global elements of T. Note that contrary to the unique morphism of codomain 1 for any object, there is no reason that global elements exist in general. However, suppose that there is a global element $x : 1 \to \mathrm{CAT}$ in our previous example. The properties of the terminal object ensure that $!_{\mathrm{CAT}} \circ x = \mathrm{id}_1$. So, by composing x with each side of the equality in (4), we get the one in (5), which goes closer to the idea we could have of a predicate calculus based on this system, as it exactly states that applying the predicate **cat** to x of type CAT raises the value true.

(5) $\mathbf{cat} \circ f \circ x = \top$

It means that for any predicate in the language we obtain in $\mathcal{T}$ both a type (an object) and a predicate term (a morphism), which are related by the Ω-axiom. This duality of monadic predicates has been discussed by Retoré [16] and more recently by Chatzikyriakidis and Luo [5], who point out that such a property could make type checking undecidable, in general. It is difficult to say whether the system presented here could bring the basis of a solution for facilitating type checking, if any; this problem will not be addressed in this paper, but should be explored in future works. Yet it is worth noticing that the two interpretations of a predicate are here related by virtue of the Ω-axiom.

At that point, we have only followed the isomorphism $\mathcal{T}(E, T) \cong \mathrm{Sub}(E)$ given by the axiom to introduce new types in our topos, but we have not specified which structure they have. By general properties of topoi, we know that $\mathrm{Sub}(E)$ is a Heyting algebra. Furthermore, a formal hierarchy can be reconstructed. Indeed, ontological inclusions can be encoded by saying that a type A is a *subtype* of B if and only if $A \cap B \cong A$, which means that any entity which satisifies the predicate P_A associated with A satisfies also the predicate P_B associated with

---

[7] This morphism is actually the character of A as a subobject of itself.

B. In this case, the properties of intersection objects ensure that there is a monomorphism from A to B. Conversely, whenever two subobjects A and B of E are linked by a monomorphism $f : A \rightarrowtail B$, then P_B can be proved to be true on A. More specifically, if P_B is the character of B as subobject of E, we want to show that $P_B \circ g \circ f = \text{true}_A$, i.e. that the diagram below commutes:

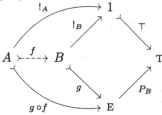

In this diagram, $g : B \rightarrowtail E$ is the subobject induced by the predicate P_B, which is a monomorphism. Considering the hypothesis that $P_B \circ g = \text{true}_B$, we just need to compose f on the right of each side of the equality, and verify that $\text{true}_A = \text{true}_B \circ f$. This last equality comes from the universal property of the terminal object: $!_B \circ f$ is a morphism $A \rightarrow 1$; but such a morphism is actually unique, so $!_B \circ f = !_A$, and then we get the desired result by composing with $\top$ on the left. Also, the morphism h given by the pullback property is such that $g \circ f = g \circ h$, which then gives us that $f = h$ because g, as monomorphism, is left-cancellative. This result can be formalised in the Property 1 below, which holds for our topos $\mathcal{T}$. It synthesises the structural condition which makes $\mathcal{T}$ a good categorical representation of a type ontology.

Property 1. A is a subtype of B in the type ontology of language if and only if there is a monomorphism from A and B in $\mathcal{T}$.

Hence we have a correspondence between subtypes as we would intend it in a type ontology, and monomorphisms in $\text{Sub}(E)$. As a result, the predicates from the language were used to build a hierarchy of types solely upon the type E, corresponding to the Montagovian general type of entities. Now E is the greatest type in the hierarchy formed by the algebra $\text{Sub}(E)$, and all of them enjoy the Ω-axiom thanks to the subobject classifier $\top$. The types formed using the operations on the algebra receive the expected interpretation: $A \cup B$ is the union type of A and B, that is the type of entities satisfying $P_A \vee P_B$; $A \cap B$ is the type of entities satisfying $P_A \wedge P_B$, $A \Rightarrow B$ is the one of entities satisfying $P_A \Rightarrow P_B$, and $\bar{A}$ the one of entities satisfying $\neg P_A$. However, it leads to a system which has arguably too many types, and we do not want to be bound to use them all. We will thus examine in the following how the types required for a semantic calculus could be restricted.

3.2   From Type Hierarchy to an Ontology of Types

We have seen that every first-order monadic predicate in the language defines its own type in the hierarchy $\text{Sub}(E)$, so that the algebra has at least as many types as such predicates. Yet the vocabulary of a given language is finite, and so are the "base types" directly created from language predicates. However, building new types on those basic ones by the means of negation, union, intersection, and implication operators quickly leads to a combinatorial explosion of the number of types. One possible solution to avoid the troubles of a direct implementation of our system would be to select only a finite (and possibly small) substructure of Sub(E) to be used with all the terms of the future λ-calculus. This idea arises from the fact that the whole type hierarchy cannot be an ontology, as argued for instance in [24]. Indeed, ontological types can be understood as types which divide the class of entities at a general level: types arising from predicates like 'cat', 'unicorn' or 'fork' are too specific to divide the world in an important way, contrary to distinctions like physical or abstract, or animate or not.

This actually matches Sommers' view on categories: they are defined from predicates not by truth, but by span. When we said that this entity is a cat, neither the language nor the ontologist bother with whether this entity is actually a cat, but only with whether this entity has the required properties to wonder whether it is a cat or not. Thus ontological types would be the way to encode the presuppositions that are made whenever we apply a predicate to a given entity. Then, considering our current type system, is there a way to retrieve Sommers' ontological categories? It is actually quite easy when we compare what we have done so far to the construction proposed by Sommers. As presented in the introduction, Sommers built his categories from predicates, and then defined types of predicates which define the same categories; we have similarly built our types from predicates. The only difference between Sommers' categories and our types is that entities in categories are *spanned* by the defining predicate, whereas entities in our types are *true* for the defining predicate.

Moving from truth to span can be done directly by following the definition of what Sommers named a predicate: as detailed in [21], the class $|P|$ of entities spanned by a predicate P is exactly the class of entities which are P or un-P, that is, which satisfy either P or $\neg P$. In our topos, we can get such classes by considering only predicates of the form $P \vee \neg P$, and the types we obtain are of the form $C(A) \doteq A \cup \bar{A}$. Hence an important property that should be imposed to our topos if we want to get the same ontology as Sommers is to be non-classical. Indeed, if our topos is classical, then every subobject algebra, including Sub(E), becomes a Boolean algebra, which means that for every $A \in$ Sub(E) we have $A \cup \bar{A} \cong$ E. It is interesting to notice here that as a result, if we constrain T to be a classical topos and if we apply the procedure described here, all first-order monadic predicates define the type E, so that a pure Montagovian system is retrieved. It is also important to point out that making our topos non-classical does not necessarily mean that we are adding one or several new truth values in our system, in the sense of morphisms $1 \rightarrow$ T: the topos T may still be bivalent, even if T does not behave like a two-element set. Rather, we are moving from classical logic to intuitionistic logic, where the law of excluded middle no longer applies: the application of a predicate to an entity may not raise a truth value, and if it happens we are in the situation of a category mistake in the sense of Ryle. Hence our type system based on a non-classical topos is able to capture the behaviour of Sommers' ontological categories. This leads to another property which holds for our topos T:

Property 2. Sommers' ontological categories are definable in $\mathcal{T}$ if and only if $\mathcal{T}$ is non-classical.

As recalled above, Sommers made use of his categories to gather predicates which define the same categories under the same A-types. In our type system, this simply corresponds to refining the domain of the predicate morphisms to their span. For instance, consider the predicates **cat** and **dog** and suppose that their spans are the same, that is, $C(\text{CAT}) \cong C(\text{DOG})$. To fix the ideas, let the type of their span (up to isomorphism) be ANIMATE.[8] Then, we would have to replace **cat** and **dog** by restricted predicates **cat'** and **dog'** of type ANIMATE $\rightarrow$ T. Such a transformation is enabled by composing with the corresponding monomorphism: whenever $f : A \rightarrowtail \text{E}$ is a subobject of E, we can send any predicate $g : \text{E} \rightarrow \text{T}$ to $g \circ f : A \rightarrow \text{T}$.[9] Therefore we can apply such a function to **cat** and **dog** because ANIMATE is a subobject of E, and the results are the desired **cat'** and **dog'**. Henceforth we will only consider predicates with domains refined to their span, and ontological types—that is, span types in question—for typing entities. Thus the size of an implementation of a calculus based on this type system can be reduced since—following Sommers' arguments—our new ontological system is a skeleton of the complete hierarchy we had first, with far fewer types needed.

Yet at this point we are unsure about the structure of our ontology. In general in a topos, whenever two subobjects A and B are linked by a monomorphism then there is a monomorphism from $\bar{B}$ to $\bar{A}$,[10] and we can say nothing about the relationship between $C(A)$ and $C(B)$. In our case however, Sommers brought a solution by formulating his structural principle. It was intended to reflect the "ontological behaviour" of the language, and claims that two overlapping categories are included one in the other. Consider for instance the types DOG, CANID and ANIMATE, with quite obvious subtyping relations between them. Even if DOG is a strict subtype of CANID, it seems reasonable to assume that their respective spans are actually the same, i.e. $C(\text{DOG}) \cong C(\text{CANID})$, because a term that would span one but not the other could hardly be found. However, as

---

[8] Whether or not the common span of 'cat' and 'dog' is really animate entities could be debated, in particular with some examples as in (i) where 'rock' seems also to belong to the span of 'dog':

(i)   This is not a dog but just a rock.

  However, the main idea to keep is that the spans of the two predicates are probably the same, as it does not seem absurd to say that anything that is not a dog *could* be a cat or not, and conversely.

[9] This transformation corresponds to a function $\mathcal{T}(\text{E}, \text{T}) \rightarrow \mathcal{T}(A, \text{T})$ in Set, which is a specific case of application of the contravariant hom-functor $\mathcal{T}(-, \text{T})$.

[10] This is actually a well-known property for Hilbert algebras, but it applies here as a Heyting algebra is a particular case of Hilbert algebra. The existence of a monomorphism between two subobjects A and B corresponds to the natural order of those algebras: if we note $A \leq B$ when such a monomorphism exists, then $A \leq B$ iff $(E \cap A) \leq B$ iff $E \leq (A \Rrightarrow B)$, which is equivalent to $E \cong A \Rrightarrow B$ as expected for a natural order.

previously supposed, the ontological category $C(\textsc{dog})$ is isomorphic to ANIMATE, which means that there is a monomorphism from $C(\textsc{dog})$ to $C(\textsc{animate})$ which is not an isomorphism: in other words, the category $|\textbf{dog}|$ is strictly included within the category $|\textbf{animate}|$. If all predicates obey this principle we retrieve a hierarchical structure in our ontology, where the passage from general types to ontological ones has made some parts of the hierarchy "collapse" into the same span type, while other subtyping relations are preserved.

The construction presented here is admittedly rather abstract and gives no clue about how to build such an ontology in practice. In [16], Retoré provides an interesting discussion on what should be the base types of a semantic type system, and quickly reviews the different set of types that have been proposed so far. As he states himself, the choice of such a set depends notably "on one's philosophical convictions", and this problem will not be solved here. However, the type system presented in this paper can accomodate with any proposition, because it actually generates the greatest number of types possible by default, and as shown in this section this overgeneration does not preclude to refine the set of types used in practice as long as it keeps the structure of a sub-hierarchy. Thus the pure Montagovian system is retrieved when taking the span types in a classical topos, Chatzikyriakidis and Luo's proposition of common nouns as types [5] is captured quite easily from our original type construction, and intermediate sets as Sommers' and others' can also be used in such a framework.

3.3   A Short Account of Dot Objects

One of the main ideas of Asher [1] for his categorical model of TCL was to propose a type-theoretic account for the so-called *dot objects* [2,15]. Dot objects are lexical units which show the property of inherent polysemy: they can appear in contexts that are generally contradictory in terms of type requirement. A classical exemple of dot object is 'book', which can be treated as a physical object (6a), or as an informational content (6b), even though physical objects and informational content have distinct, non-overlapping spans.

(6)   a.   Mary picked up the book.
　　　 b.   John didn't understand the book.

Hence dot objects are entities with several separate aspects, that is, several types. In Pustejovsky's and Asher's works, those objects are given a *dot type*, that is a complex type structure where every type aspect is represented. In the case of 'book', if we call P the type of physical objects and I the type of informational contents,[11] the entry for 'book' would receive the type P • I.

The major concern about dot types is to understand where they should be placed in the hierarchy. It is commonly admitted that dot types cannot be inter-section types, as the intersection of two uncompatible types is naturally supposed to be empty (cf. [1, Chap. 5] for discussion). Another hypothesis is to consider

---

[11] In the remainder of this paper both types will be assumed to be ontological.

such types to be pairs of types. As pointed out by Asher, this can be an interesting solution provided that the different aspects of a dot object are kept separate, instead of having the transformation only on the type side—that is, we do not want to consider a dot type to be a direct subtype of its components. To sum up, he introduces in his categorical model dot types as objects with "aspect projections" to some pullback objects.[12] Those pullback are more precisely defined from the relation between aspects of the dot object, lifted to power objects. For 'book' of type $P \bullet I$, let $ex : I \to \mathscr{P}(P)$ and $in : P \to \mathscr{P}(I)$ be those lifted relations: then, $P \bullet I$ has projections to the pullback objects of the diagrams $I \xrightarrow{ex} \mathscr{P}(P) \xleftarrow{id} \mathscr{P}(P)$ and $\mathscr{P}(I) \xrightarrow{id} \mathscr{P}(I) \xleftarrow{in} P$.

As explained by Pustejovsky [15], the relation between aspects is part of the definition of a dot object, which means that for a given pair of types several different relations—and consequently several different dot objects—can be defined. The $\bullet$ operator only says that a relation exists, but does not provide it explicitly. Pustejovsky proposed to define several dot operators $\bullet_{R_1}, \ldots, \bullet_{R_n}$, one for each relation $R_1, \ldots, R_n$, for a proper account. In our topos, we can follow this idea by introducing the relations explicitly as subobjects of a product. Thus, the type of 'book' would be defined by an object BOOK equipped with a monomorphism $f : \text{BOOK} \rightarrowtail P \times I$. Such a definition is actually equivalent to Asher's proposition: indeed, the properties of power objects (which always exist in a topos) state that the the relation object BOOK implies the existence of the two lifted morphisms $book_P : P \to \mathscr{P}(I)$ and $book_I : I \to \mathscr{P}(P)$. Then the two projections proposed by Asher can be retrieved from the following compositions:

$$
\begin{array}{ccc}
 & \text{BOOK} & \\
 & \Big\uparrow f & \\
P \times \mathscr{P}(I) \xleftarrow{\langle \pi_1, book_P \circ \pi_1 \rangle} & P \times I & \xrightarrow{\langle book_I \circ \pi_2, \pi_2 \rangle} \mathscr{P}(P) \times I
\end{array}
$$

where $\pi_1 : P \times I \to P$ and $\pi_2 : P \times I \to I$ are the canonical projections of the product. Note that the morphisms above are not exactly the ones from Asher, as their codomains are products instead of pullbacks, that is in this case restricted products w.r.t. the satisfaction of the relation BOOK; but Asher's can be obtained easily from those by *epi-monic factorisations*, which are always possible in a topos (see [9] for details).

We have therefore a way for representing dot types without using the dot notation, which thus permits placement of the relation between the different aspects at the centre of the definition of such a type. Being a dot type in T is equivalent to being a subobject of a product of types from Sub(E). When browsing Sommers' theory, it is difficult not to make the connection between dot objects and heterotypical entities, because of this common ability to show multiple types according to the context. Actually, some arguments pointing out that the former may be a particular case of the latter can be given. Besides this

---

[12] There is actually more subtleties in his construction, but they will not be detailled here due to lack of space. The whole reasoning can be found in [1].

multiple typing property, it is interesting to notice that in general, relations such as BOOK do *not* belong to Sub(E), because the product of two subobjects is not itself a subobject. The projection morphisms from a product to its components has no reason to be a monomorphism, unless all components but one are terminal objects. Moreover, in general their compositions with subobjects of the product do not create monomorphisms either, which is also an expected behaviour: in the case of 'book', there are obviously many copies of the same book as well as copies compiling several books in one physical object, which means in a set-theoretic acceptance (for fixing ideas) several pairs with same image through the projections. Thus if our type ontology is included in Sub(E), then dot objects cannot be part of it—and this exactly is how heterotypicals behave. A last note we should make here is that relation types like BOOK, as a consequence of their definition, are not ontological; and following Sommers' philosophy the product type P ● I is not ontological either. But that does not mean that we cannot use those types in practice. Rather, this should incite us to treat such types for what they really describe: relations between ontological types.

4   Related Works and Future Perspectives

Several works which have been great sources of inspiration for the topos type system presented here have been mentioned throughout this paper. It will thus not be surprising that connections to those works could be made. As shown above, the natural definition of types from language predicates share many common points with the proposition of common nouns as types advocated by Luo and Chatzikyriakidis [5]. Moreover, the organisation of types in a Heyting algebra of subobjects allows parallels with Luo's coercive subtyping [12], as categorical properties make compositions coercive rather than subsumptive: if we have in our topos T defined above a predicate $P : A \to T$ to be applied to an entity x of type B (that is, a global element $x : 1 \to B$) with B subtype of A, then the only way to compose both morphisms in T a priori is to use the monomorphism f between B and A. This leads to the morphism $P \circ f \circ x$, where f is a subtyping coercion in the sense of Luo. It is also worth noticing that his complete theory partially originates from Martin-Löf's intuitionistic type theory: as natural basis for intuitionistic logic, topoi could be useful for a categorical-based account of this approach.

The connection between the initial proposition of Asher [1] for TCL and the present work is also unsurprising, as the latter is mainly a deeper look into the properties of topoi and their consequenses on type systems, heavily inspired by the former. However, nothing has been said about the contravariance problem for subtyping between monadic predicates in this paper. A categorical-based argument can be given to advocate for the existence of a covariant subtyping of first-order arrow types, and more developments on this idea could be given in the future. The work of Retoré and Mery on $\Lambda T y_n$ (see [16]), a framework based on Girard's system F, has also been mentioned here, but further investigations seem to be necessary to determine whether the present work could be extended to a semantic model of their system.

As for the question of type ontologies, it has been argued above that the topos-based system described in these pages can be adapted to any set of base types, depending on one's philosophical convictions on that matter, including the traditional Montagovian system. It has also been shown that this type system shares interesting and welcome similarities with the theory of Fred Sommers. The ontological tree he proposed [20] can be reconstructed as a substructure of the algebra $Sub(E)$ in a natural way, and the topos even seems to give a faithful account of the case of heterotypical composites as lying outside the tree structure. Although Sommers' theory has been questioned extensively in the 70s, it seems to have been somewhat forgotten since. However his view on ontologies could have useful applications in the fields of formal and lexical semantics. This has been also recently advocated by Saba [18], and the present work might serve as a logical basis for such approaches. It is also worth noticing that Sommers, as follower of the theory of meaning-in-use, proposed a concrete way of building his ontology from language. An actual hierarchy could thus be obtained by implementing and running his method on corpora.

This paper presented a sketch of what would be a type system based on topoi, introducing a specific instance of a topos which shows how to construct various kind of types ("classical", ontological, heterotypical), and how to organise them in a hierarchical structure able to produce the type systems usually assumed in formal and lexical semantics. Moreover, two main properties that have to be satisfied by the topos have been drawn: monomorphisms should represent subtyping relations, and the topos should be non-classical. As it has been pointed out several times in these pages, such a categorical type system should naturally lead to a typed λ-calculus, using objects of the topos as types and morphisms as terms. The formal construction and the properties of such a calculus are still to be explored, and should therefore constitute the general outline of future work on this subject. More particularly, the question of how this system can be improved in order to give new tools for a fine-grained account of type shifts, coercion and copredication phenomena will be investigated at some point.

References

1. Asher, N.: Lexical Meaning in Context: A Web of Words. Cambridge University Press, Cambridge (2011)
2. Asher, N., Pustejovsky, J.: A type composition logic for generative lexicon. J. Cogn. Sci. **7**(1), 1–38 (2006)
3. Berry, G.: Some syntactic and categorical constructions of lambda-calculus models, RR-0080. Inria (1981)
4. Brown, R.: A First Language: The Early Stages. Harvard University Press, Cambridge (1973)
5. Chatzikyriakidis, S., Luo, Z.: On the interpretation of common nouns: types versus predicates. In: Chatzikyriakidis, S., Luo, Z. (eds.) Modern Perspectives in Type-Theoretical Semantics. SLP, vol. 98, pp. 43–70. Springer, Cham (2017). https://doi.org/10.1007/978-3-319-50422-3_3
6. Chomsky, N.: Some methodological remarks on generative grammar. Word **17**(2), 219–239 (1961)

7. Church, A.: A formulation of the simple theory of types. J. Symb. Logic **5**(2), 56–68 (1940)
8. Geeraerts, D.: Prototype theory. Linguistics **27**(4), 587–612 (1989)
9. Goldblatt, R.: Topoi: The Categorial Analysis of Logic. Studies in Logic and the Foundations of Mathematics, vol. 98. North-Holland Publishings, Amsterdam (1979)
10. Kiefer, F.: Some semantic relations in natural language. In: Josselson, H.H. (ed.) Proceedings of the Conference on Computer-related Semantic Analysis, pp. VII/1–23. Wayne State University, Detroit (1966)
11. La Palme Reyes, M., Macnamara, J., Reyes, G.E.: Reference, kinds and predicates. In: Macnamara, J., Reyes, G.E. (eds.) The Logical Foundations of Cognition, Vancouver Studies in Cognitive Science, vol. 4, pp. 91–143. Oxford University Press, Oxford (1994)
12. Luo, Z.: Type-theoretical semantics with coercive subtyping. In: Li, N., Lutz, D. (eds.) Proceedings of SALT 20, vol. 20, pp. 38–56 (2010)
13. Miller, G.A.: Nouns in WordNet. In: Fellbaum, C. (ed.) WordNet: An Electronic Lexical Database, pp. 23–46. The MIT Press, Cambridge (1998)
14. Montague, R.: The proper treatment of quantification in ordinary english. In: Suppes, P., Moravcsik, J., Hintikka, J. (eds.) Approaches to Natural Language, pp. 221–242. Springer, Dordrecht (1973). https://doi.org/10.1007/978-94-010-2506-5_10
15. Pustejovsky, J.: The semantics of lexical underspecification. Folia Linguistica **32**(3–4), 323–348 (1998)
16. Retoré, C.: The montagovian generative lexicon ΛTy_n: a type theoretical framework for natural language semantics. In: Matthes, R., Schubert, A. (eds.) Proceedings of the 19th International Conference on Types for Proofs and Programs. LIPICS, vol. 26, pp. 202–229 (2014)
17. Rosch, E.H.: Natural categories. Cogn. Psychol. **4**, 328–350 (1973)
18. Saba, W.S.: Logical semantics and commonsense knowledge: where did we go wrong, and how to go forward, again (2018). arXiv preprint
19. Seely, R.A.G.: Categorical semantics for higher order polymorphic lambda calculus. J. Symb. Logic **52**(4), 969–989 (1987)
20. Sommers, F.: The ordinary language tree. Mind **68**(2), 160–185 (1959)
21. Sommers, F.: Type and ontology. Philos. Rev. **72**(3), 327–363 (1963)
22. Sommers, F.: Structural ontology. Philosophia **1**(1–2), 21–42 (1971)
23. Suzman, J.: The ordinary language lattice. Mind **81**(3), 434–436 (1972)
24. Westerhoff, J.: The construction of ontological categories. Australas. J. Philos. **82**(4), 595–620 (2004)
25. Wittgenstein, L.: Philosophical Investigations. Macmillan, New York (1953)

Structure Sensitive Tier Projection: Applications and Formal Properties

Aniello De Santo$^{(\boxtimes)}$ and Thomas Graf

Department of Linguistics, Stony Brook University, Stony Brook, USA
aniello.desanto@stonybrook.edu, mail@thomasgraf.net

Abstract. The subregular approach has revealed that the phonological surface patterns found in natural language are much simpler than previously assumed. Most patterns belong to the subregular class of tier-based strictly local languages (TSL), which characterizes them as the combination of a strictly local dependency with a tier-projection mechanism that masks out irrelevant segments. Some non-TSL patterns have been pointed out in the literature, though. We show that these outliers can be captured by rendering the tier projection mechanism sensitive to the surrounding structure. We focus on a specific instance of these *structure-sensitive TSL* languages: input-local TSL (ITSL), in which the tier projection may distinguish between identical segments that occur in different local contexts in the input string. This generalization of TSL establishes a tight link between tier-based language classes and ISL transductions, and is motivated by several natural language phenomena.

Keywords: Subregular hypothesis · TSL · Phonotactics · Input strictly local functions · Generative capacity

1 Introduction

The subregular hypothesis ([16] and references therein) posits that every language's set of phonologically well-formed surface strings—its phonotactic patterns—belongs to a proper subclass of the regular languages. The class of tier-based strictly local languages (TSL) has been of particular interest in this respect [17]. TSL is inspired by autosegmental phonology [12] and combines two components: (I) an n-gram based mechanism to enforce local constraints on adjacent segments, and (II) a tier projection mechanism that "masks out" irrelevant parts of the string. Long-distance dependencies are thus reanalyzed as local dependencies over strings with masked out segments.

While TSL covers a wide range of data, recent literature has reported several instances of complex phenomena—from Samala sibilant harmony to unbounded tone plateauing—that cannot be characterized in these terms [14,15,24, a.o.]. We argue that all these counterexamples can be accounted for by extending the tier projection mechanism. We redefine TSL as a cascade of three string transductions, one of which is the tier projection mechanism. In standard TSL, the

© Springer-Verlag GmbH Germany, part of Springer Nature 2019
R. Bernardi et al. (Eds.): FG 2019, LNCS 11668, pp. 35–50, 2019.
https://doi.org/10.1007/978-3-662-59648-7_3

tier projection is an input strictly local function of locality 1 (1-ISL) in the sense of Chandlee [5, Definition 4]. By allowing for more complex string transductions, one obtains the much more powerful class of *structure sensitive TSL* (SS-TSL). Within this wide range of options, we focus on the natural generalization from 1-ISL to n-ISL. This means that projection of a segment s does not merely depend on s alone but may also consider the locally bounded context $u_1 \cdots u_m \text{ - } v_1 \cdots v_n$ in which s occurs. The resulting class of *input tier-based strictly local* (ITSL) languages greatly expands the empirical coverage of TSL while retaining essential formal properties.

The paper is structured as follows. Section 2 introduces mathematical notation that is essential for studying subregular languages. The fundamental properties of strictly local (SL) and tier-based strictly local (TSL) languages are presented in Sect. 3. There, we also introduce the first major innovation of this paper, the generalization from standard TSL to SS-TSL. We then define ITSL, the most natural subclass of SS-TSL. Section 4 studies the formal properties of ITSL, and relates it to the rest of the subregular hierarchy. We then expand on this with results on the intersection closures of TSL and ITSL, respectively (Sect. 5). Finally, Sect. 6 discusses the implications of these results for learnability.

2   Preliminaries

This paper discusses TSL and our generalization of its projection function. As we compare the resulting new languages to several subregular classes besides TSL, a fair amount of mathematical machinery is required. We assume familiarity with set notation on the reader's part.

Given a finite alphabet Σ, Σ^* is the set of all possible finite strings of symbols drawn from Σ. A language L is a subset of Σ^*. The concatenation of two languages $L_1 L_2 = \{uv : u \in L_1 \text{ and } v \in L_2\}$. For every string w and every nonempty string u, $|w|$ denotes the length of the string, $|w|_u$ denotes the number of occurrences of u in w, and ε is the unique empty string. Left and right string boundaries are marked by $\rtimes, \ltimes \notin \Sigma$ respectively.

A string u is a *k-factor* of a string w iff $\exists x, y \in \Sigma^*$ such that $w = xuy$ and $|u| = k$. The function F_k maps words to the set of k-factors within them:

$$F_k(w) := \{u : u \text{ is a } k\text{-factor of } w \text{ if } |w| \geq k, \text{ else } u = w\}$$

For example, $F_2(aab) = \{aa, ab\}$. The domain of F_k is generalized to languages $L \subseteq \Sigma^*$ in the usual way: $F_k(L) = \bigcup_{w \in L} F_k(w)$. We also consider the function which counts k- factors up to some threshold t.

$$F_{k,t}(w) := \{(u, n) : u \text{ is a } k\text{-factor of } w \text{ and } n = \min(|w|_u, t)\}$$

For example $F_{2,5}(aaaaab) = \{(aa, 4), (ab, 1)\}$, but $F_{2,3}(aaaaab) = \{(aa, 3), (ab, 1)\}$.

In order to simplify some proofs, we rely on first-order logic characterizations of certain string languages and string-to-string mappings. We allow standard

Boolean connectives ($\wedge$, $\vee$, $\neg$, $\rightarrow$), and first-order quantification ($\exists$, $\forall$) over individuals. We let $x \prec y$ denote *precedence*, $x \approx y$ denote *identity*, and x, y denote variables ranging over positions in a finite string $w \in \Sigma^*$. Note that $\prec$ is a strict total order.

The remaining logical connectives are obtained from the given ones in the standard fashion, and brackets may be dropped where convenient. For example, *immediate precedence* is defined as $x \vartriangleleft y \Leftrightarrow x \prec y \wedge \neg \exists z[x \prec z \wedge z \prec y]$. We add a dedicated predicate for each label $\sigma \in \Sigma$ we wish to use: $\sigma(x)$ holds iff x is labelled σ, where x is a position in w.

Classical results on definability of strings represented as finite first-order structures are then used [26]. If $\Sigma = \{\sigma_1, \ldots, \sigma_n\}$, then a string $w \in \Sigma^*$ can be represented as a structure M_w in the signature$(\sigma_1(\cdot), \ldots, \sigma_n(\cdot), \prec)$. If φ is a logical formula without any free variables, we use $L(\varphi) = \{w \in \Sigma^* \mid M_w \text{ satisfies } \varphi\}$ as the stringset extension of φ .

3   Structure-Sensitive TSL Languages

There is a rich literature exploring the subclasses that the regular languages can be divided into [4,9,27,32, a.o.]. Among these *subregular* classes, *tier-based strictly local* languages (TSL; [17]) have received particular attention due to their ability to provide natural descriptions of phonological well-formedness conditions (see also [13,19,29]). TSL extends the class of *strictly local* languages (SL) with a tier projection mechanism that renders non-local dependencies in a string local over tiers. The projection mechanism is very limited though, as it only considers a segment's label but not its structural context. This is too restrictive for phonology, which is why we extend TSL to a class of languages sensitive to structural information: TSL where tier projection can take local information into account.

3.1   Strictly Local and Tier-Based Strictly Local Languages

SL is the class of languages that can be described in terms of a finite number of forbidden substrings. Intuitively, SL languages describe patterns which depend solely on the relation between a bounded number of consecutive symbols in a string—there are no long-distance dependencies.

Definition 1 (SL). *A language L is* strictly k-local *(SL_k) iff there exists a finite set $S \subseteq F_k(\rtimes^{k-1}\Sigma^*\ltimes^{k-1})$ such that*

$$L = \{w \in \Sigma^* : F_k(\rtimes^{k-1}w\ltimes^{k-1}) \cap S = \emptyset\}.$$

We also call S a strictly k-local grammar, and we also use $L(S)$ to indicate the language generated by S. A language L is strictly local iff it is SL_k for some $k \in \mathbb{N}$.

For example, $(ab)^n$ is a strictly 2-local language over alphabet $\{a, b\}$ because it is generated by the grammar $G := \{\rtimes b, bb, aa, a\ltimes\}$.[1]

Even though this paper is concerned with extensions of SL, many of our proofs make use of a particular characterization of SL in terms of k-local suffix substitution closure [30].

Definition 2 (Suffix Substitution Closure). *For any $k \geq 1$, a language L satisfies k-local suffix substitution closure iff for all strings u_1, v_1, u_2, v_2, for any string x of length $k - 1$ if both $u_1 \cdot x \cdot v_1 \in L$ and $u_2 \cdot x \cdot v_2 \in L$, then $u_1 \cdot x \cdot v_2 \in L$.*

Theorem 1. *A language is SL_k iff it satisfies k-local suffix substitution closure.*

The language $L := a^*ba^*$, for example, is not SL because for any k we can pick two strings $a^m ba^k \in L$ and $a^k ba^n \in L$ and recombine them into $a^m ba^k ba^n \notin L$. However, this language is TSL.

TSL is an extension of SL where k-local constraints only apply to elements of a tier $T \subseteq \Sigma$. An erasing function (also called projection function) is introduced to delete all symbols that are not in T. Given some $\sigma \in \Sigma$, the erasing function $E_T : \Sigma \to \Sigma \cup \{\varepsilon\}$ maps σ to itself if $\sigma \in T$ and to *mptystring* otherwise.

$$E_T(\sigma) := \begin{cases} \sigma & \text{if } \sigma \in T \\ \varepsilon & \text{otherwise} \end{cases}$$

We extend E_T from symbols to strings in the usual pointwise fashion.

Definition 3 (TSL). *A language L is tier-based strictly k-local (TSL$_k$) iff there exists a tier $T \subseteq \Sigma$ and a finite set $S \subseteq F_k(\rtimes^{k-1} T^* \ltimes^{k-1})$ such that*

$$L = \{w \in \Sigma^* : F_k(\rtimes^{k-1} E_T(w)\ltimes^{k-1}) \cap S = \emptyset\}$$

We also call S the set of forbidden k-factors on tier T, and $\langle S, T \rangle$ is a TSL$_k$ grammar.

As can be gleaned from Definition 3, a language L is TSL iff it is strictly k-local on tier T for some $T \subseteq \Sigma$ and $k \in \mathbb{N}$. This will be important for many proofs.

For a concrete example, consider once more $L := a^*ba^*$ such that $aba, aabaa, aaaba \in L$ but $abaabaa, ababaa \notin L$. This language is generated by the TSL$_2$ grammar $\langle \{\rtimes\ltimes, bb\}, \{b\} \rangle$ over $\Sigma = \{a, b\}$, which bans every string whose tier is empty (no b) or contains more than one b.

---

[1] A comment regarding edge markers. For S to be k-local, it needs to contain only factors of length k. Thus, strings are augmented with enough edge markers to ensure that this requirement is satisfied. However, it is often convenient to shorten the k-factors in the definition of strictly k-local grammars and write down only one instance of each edge marker. with the implicit understanding that it must be augmented to the correct amount. So $\rtimes \rtimes a$ is truncated to $\rtimes a$. We adopt this simpler notation throughout the paper, unless required to make a definition clearer.

3.2   Insufficiency of TSL

While TSL enjoys wide empirical coverage in phonology, some non-TSL phenomena have been pointed out in the literature [14,15,24]. As a concrete example, consider the case of sibilant harmony in Samala, where an unbounded dependency can override a local one (see [2] for the original data set and [24] for a subregular analysis). Samala displays sibilant harmony such that [s] and [ʃ] may not co-occur anywhere within the same word (cf. Ex. (1a)). There is also a ban against string-adjacent [st], [sn], [sl], which is resolved by dissimilation of [s] to [ʃ] (cf. Ex. (2a) and (2b)). However, dissimilation is blocked if the result would violate sibilant harmony. Thus /sn/ surfaces as [ʃn] unless the word contains [s] somewhere to the right, in which case it is realized as [s] (cf. Ex. (2a) and (3a)).

(1)   a. /k-su-ʃojin/ → [kʃuʃojin]

(2)   a. /s-ni?/ → [ʃni?]

     b. /s-ni?/ → *[sni?]

(3)   a. /s-net-us/ → [snetus]

     b. /s-net-us/ → *[ʃnetus]

This pattern is not TSL. Pick some sufficiently large m and consider the strings [sne(ne)^{m}tus] and [ne(ne)^{m}tus], which are well-formed according to the generalization above. In stark contrast, the minimally different [sne(ne)^{m}tu] is ill-formed. In order to regulate this dependency, we need a TSL grammar whose tier contains at least [s] and [n]. But then the tiers of these three strings are of the form snn^{m}s, nn^{m}s, and snn^m, respectively. By suffix substitution closure, it is impossible for an SL grammar to allow the former two while forbidding the latter. But if the tier language is not SL, the original language is not TSL, either. Note that projecting additional symbols does not change anything with respect to suffix substitution closure, so the problem is independent of what subset of Σ one chooses as the tier alphabet.

The central shortcoming of TSL is that it only provides a choice between projecting no instance of [n], which is obviously insufficient, and projecting every instance of [n], which renders the dependency between sibilants non-local over tiers. But suppose that one could instead modify the projection function such that an [n] is projected iff it is immediately preceded by a sibilant. Then [sne(ne)^{m}tus] and [ne(ne)^{m}tus] have the tiers sns and s, whereas [sne(ne)^{m}tu] has the tier sn. An SL$_3$ grammar can easily distinguish between these, permitting the former two but not the latter. Such a modified version of TSL will also be able to block [snetu] while allowing for [senetu] as their respective tiers are sn and s. Apart from this Samala example, reported non-TSL patterns that can be accounted for by inspecting the local context of a segment before projecting it include nasal harmony in Yaka [33], unbounded stress of Classical Arabic (see [3] and references therein), Korean vowel harmony [14], and cases of unbounded tone plateauing [20, a.o.].

More recently, other patterns have been reported for which it seems to be necessary to extend TSL projections to consider more than just local contexts in the input string. Mayer and Major [23], based on a suggestion by Graf (p.c.), make tier-projection sensitive to preceding segments on the tier in order to capture backness harmony in Uyghur. Graf and Mayer [15] analyze Sanskrit retroflexion in terms of an even more general class whose projection function considers the local contexts in both the input string and the already constructed tier.

Crucially, all these extensions allow the erasing function E_T to consider additional structural factors. We call all languages in which the projection function has been extended along these lines *structure-sensitive* TSL. This is a very loosely defined class, but as we explain next the idea can be made more precise by viewing TSL-like grammars as a cascade of three string transductions.

3.3   TSL as the Composition of Three Transductions

For every TSL grammar $G := \langle S, T \rangle$, one can construct a sequence of transductions that generates exactly the same string language:

1. The *projection transduction* E_T rewrites every symbol $s \in T$ as s and deletes every $s' \notin T$.
2. The *grammar transduction* id_S is the identity function over $L(S)$.
3. The *filler transduction* F_T is the inverse of E_T.

Their composition $E_T \circ \mathrm{id}_S \circ F_T$ is a partial, non-deterministic finite-state transduction. The image of Σ^* under this transduction is exactly $L(G)$. All the recent extensions of TSL keep id_S the same, but they change the nature of E_T (and hence F_T). Without further limitations on E_T, every recursively enumerable string language can be generated this way. But from a linguistic perspective, this is immaterial as only very limited kinds of SS-TSL have been proposed. These classes generalize E_T to ISL or OSL functions as originally defined in [5]. We only consider the former here and leave the latter for future work.

3.4   Input-Sensitive TSL

Adding input-sensitivity to TSL only requires a minor change to the definition of E_T. In order to simplify the exposition later on, we take inspiration from [7] and define ISL projections in terms of local contexts.

Definition 4 (Contexts). *A k-context c over alphabet Σ is a triple $\langle \sigma, u, v \rangle$ such that $\sigma \in \Sigma$, $u, v \in \Sigma^*$ and $|u| + |v| \leq k$. A k-context set is a finite set of k-contexts.*

Definition 5 (ISL Projection). *Let C be a k-context set over Σ (where Σ is an arbitrary alphabet also containing edge-markers). Then the input strictly k-local (ISL-k) tier projection π_C maps every $s \in \Sigma^*$ to $\pi'_C(\rtimes^{k-1}, s\ltimes^{k-1})$, where $\pi'_C(u, \sigma v)$ is defined as follows, given $\sigma \in \Sigma \cup \{\varepsilon\}$ and $u, v \in \Sigma^*$:*

$$
\begin{array}{ll}
\varepsilon & \text{if } \sigma a v = \varepsilon, \\
\sigma \pi'_C(u\sigma, v) & \text{if } \langle \sigma, u, v \rangle \in C, \\
\pi'_C(u\sigma, v) & \text{otherwise.}
\end{array}
$$

Note that an ISL-1 tier projection only determines projection of σ based on σ itself, just like E_T does for TSL. This shows that ISL-k-tier projections are a natural generalization of E_T even though they are no longer defined in terms of some $T \subseteq \Sigma$. The definition of ITSL languages then closely mirrors the one for TSL.

Definition 6 (ITSL). *A language L is m-input local k-TSL (m-ITSL$_k$) iff there exists an m-context set C and a finite set $S \subseteq \Sigma^k$ such that*

$$L = \{w \in \Sigma^* : F_k(\rtimes^{k-1}\pi_C(w)\ltimes^{k-1}) \cap S = \emptyset\}.$$

A language is input-local TSL *(ITSL) iff it is m-ITSL$_k$ for some $k, m \geq 0$. We call $\langle S, C \rangle$ an ITSL grammar.*

Let us return to the interaction of local dissimilation and non-local harmony in Samala. This process can be handled by an 2-ITSL$_3$ grammar $\langle S, C \rangle$ with

- $S := \{s\int, \int s, snx\}$ where $x \in \{\Sigma - s\}$,
- C contains all of the following contexts, and only those:
  - $\langle s, \varepsilon, \varepsilon \rangle$
  - $\langle S, \varepsilon, \varepsilon \rangle$
  - $\langle n, s, \varepsilon \rangle$

Since this phenomenon could not be handled with TSL, ITSL properly extends TSL.

Theorem 2. *TSL $\subsetneq$ ITSL*

For the sake of rigor, we also provide a formal proof.

Proof. TSL $\subseteq$ ITSL is trivial. Now consider the language $L = a\{a,b\}^*b \cup b\{a,b\}^*a$ over alphabet $\Sigma = \{a,b\}$. It is generated by the 2-ITSL$_2$ grammar $\langle S, C \rangle$ with $S = \{aa, bb, \rtimes\ltimes\}$ and $C := \{\langle \sigma, \rtimes, \varepsilon \rangle, \langle \sigma, \varepsilon, \ltimes \rangle \mid \sigma \in \Sigma\}$. But L is not TSL. Pick some arbitrary TSL$_k$ grammar $\langle S, T \rangle$ and strings $s := a^m b^n \in L$, $t := b^n a^o \in L$, and $u := a^m b^n a^o \notin L$ ($m, n, o > k$). These three strings witness that no matter how one chooses $T \subseteq \Sigma$, the resulting tier language is not closed under suffix substitution closure. Thus, L is not k-TSL for any k.

ITSL is clearly more powerful than TSL, but the question is how much additional power the move to ISL projections grants us. We do not want ITSL to be too powerful as it should still provide a tight characterization of the limits of natural language phonology. The next section shows that ITSL is still a very conservative extension of TSL that is subsumed by the star-free languages and largely incomparable to any other subregular classes.

4 Formal Analysis

It is known that TSL is a proper subclass of the star-free languages (SF) and is incomparable to the classes locally testable (LT), locally threshold-testable (LTT), strictly piecewise (SP), and piecewise testable (PT) [17]. In addition, TSL is not closed under intersection, union, complement, concatenation, or relabeling (this is common knowledge but has not been explicitly pointed out in the literature before). The same holds for ITSL. This is not too surprising as ITSL is a fairly minimal extension of TSL, and many of the proofs in this section are natural modifications of the corresponding proofs for TSL.

4.1 Relations to Other Subregular Classes

First we have to provide basic definitions for subregular classes we wish to compare to ITSL.

Definition 7 *(Locally t-Threshold k-Testable). A language L is locally t-threshold k-testable iff* $\exists t, k \in \mathbb{N}$ *such that* $\forall w, v \in \Sigma^*$, *if* $F_{k,t}(w) = F_{k,t}(v)$ *then* $w \in L \Leftrightarrow v \in L$.

Intuitively *locally threshold testable* (LTT) languages are those whose strings contain a restricted number of occurrences of any k-factor in a string. Practically, LTT languages can *count*, but only up to some fixed threshold t since there is a fixed finite bound on the number of positions a given grammar can distinguish. Properly included in LTT, the *locally testable* (LT) languages are *locally threshold testable* with $t = 1$.

We show that LT and ITSL are incomparable. Since TSL and LTT are known to be incomparable [17], the incomparability of LTT is an immediate corollary.

Theorem 3. *ITSL is incomparable to LT and LTT.*

Proof. That ITSL is no subset of LT or LTT follows from the fact that ITSL subsumes TSL, which is incomparable to both.

We now show that LT $\nsubseteq$ ITSL. Let L be the largest language over $\Sigma = \{a, b, c\}$ such that a string contains the substring aa only if it also contains the substring bb. This language is LT but cannot be generated by any m-ITSL$_k$ grammar G, irrespective of the choice of k and m.

Suppose G generates at least strings of the form $c^*aac^*bbc^* \in L$ and $c^*bbc^* \in L$, but not $c^*aac^* \notin L$. Then G must project both aa and bb, wherefore c^*aac^* and c^*bbc^* each license projection of aa and bb, respectively (projection of one of a or b cannot depend on the other because the number of cs between the two is unbounded). But then strings of the form $(c^*aac^*)^+bb(c^*aac^*)^+ \in L$ yield a tier language $(aa)^+bb(aa)^+$. By suffix substitution closure, G also accepts any tier of the form $(aa)^+$. Therefore, $L(G) \ni (c^*aac^*)^+ \notin L$.

Next consider the *strictly piecewise* (SP) and *piecewise testable* (PT) languages [10,28,31]. These are already known to be incomparable with SL, TSL, and LTT. For any given string w, let $P_{\leq k}(w)$ be a function that maps w to the set of subsequences up to length k in w.

Definition 8 *(Piecewise k-Testable).* *A language L is piecewise k-testable iff $\exists k \in \mathbb{N}$ such that $\forall w, v \in \Sigma^*$, if $P_{\leq k}(w) = P_{\leq k}(v)$ then $w \in L \Leftrightarrow v \in L$. A language is piecewise testable if it is piecewise k-testable for some k.*

Properly included in PT, SP languages mirror the definition of SL languages by replacing $F_k(w)$ with $P_k(w)$ in Definition 1. In short, piecewise languages are sensible to relationships between segments based on *precedence* (over arbitrary distances) rather than *adjacency* (immediate precedence).

Theorem 4 *ITSL is incomparable to SP and PT.*

Proof ITSL $\not\subseteq$ SP, PT follows from the fact that ITSL includes TSL, which is incomparable to both. In the other direction, consider the SP language L that consists of all strings over $\Sigma = \{a, b, c, d, e\}$ that do not contain the subsequences ac or bd. This language is not ITSL. In order to correctly ban both ac and bd, at least one instance of a, b, c, and d must be projected in each string. Consequently, for each symbol there must be some fixed context that triggers its projection. Assume w.l.o.g. that one of these contexts is $\langle b, u, v \rangle$. Consider the strings $s := a(e^m u b v)^n \in L$, $t := (e^m u b v)^n c \in L$, and $u := a(e^m u b v)^n c \notin L$, for sufficiently large m and n. The respective tiers are $s' := ab^n$, $t' := b^n c$, and $u' := ab^n c$. By suffix substitution closure, no SL language can contain s' and t' to the exclusion of u', wherefore L is SP (and PT) but not ITSL.

The last subregular class relevant to our discussion is SF. Multiple characterizations are known, but we will use the one in terms of first-order logic as it greatly simplifies the proof that ITSL is subsumed by SF.

Definition 9 *(Star-Free).* *Star-free (SF) languages are those that can be described by first order logic with precedence.*

Theorem 5. *ITSL $\subsetneq$ SF.*

Proof. Subsumption follows from the fact that every ITSL language can be defined in first-order logic with precedence. Proper subsumption then is a corollary of LT, PT $\subseteq$ SF together with Theorems 3 and 4.

We briefly sketch the first-order definition of ITSL. First, the successor relation $\lhd$ is defined from precedence in the usual manner. Then, for every context $c := \langle \sigma, u_1 \cdots u_m, u_{m+1} \cdots u_n \rangle$ one defines a predicate $C(x)$ as

$$\exists y_1, \ldots, y_{m+n} \Big[\sigma(x) \wedge \bigwedge_{1 \leq i < m} y_i \lhd y_{i+i} \wedge y_m \lhd x \wedge x \lhd y_{m+1} \wedge \bigwedge_{m+1 \leq i < n} y_i \lhd y_{i+i} \wedge \bigwedge_{1 \leq i \leq n} u_i(y_i) \Big]$$

The context predicates form the basis for the ITSL tier predicate

$$T(x) \Leftrightarrow \bigvee_{C \text{ is a context predicate}} C(x)$$

which in turns allows us to relativize precedence to symbols on the tier:

$$x \triangleleft_T y \Leftrightarrow T(x) \wedge T(y) \wedge x \prec y \wedge \neg \exists z [T(z) \wedge x \prec z \wedge z \prec y]$$

The set of forbidden k-factors then is just a conjunction of negative literals with $\triangleleft_T$ as the basic relation.

4.2   Closure Properties

The previous section established that ITSL is a natural generalization of TSL in the sense that it displays the same (proper) subsumption and incomparability relations with respect to other classes. We now show that this parallelism between TSL and ITSL also carries over to the standard closure properties. Just like TSL, ITSL is not closed under intersection, union, complement, concatenation, or relabeling.

We start with non-closure under intersection.

Lemma 1. *ITSL is not closed under intersection.*

Proof. Consider again the SP language L that consists of all strings over $\{a, b, c, d, e\}$ that do not contain the subsequences ac or bd. As shown in Theorem 4, this language is not ITSL. But L is the intersection of two TSL (and hence ITSL) languages L_1 and L_2 s.t. $T_1 = \{a, c\}$, $S_1 = \{ac\}$ and $T_2 = \{b, d\}$, $S_2 = \{bd\}$. Thus closure under intersection does not hold.

Lemma 2. *ITSL is not closed under concatenation.*

Proof. Let L be the union of $ab \{a, b, c\}^* a$ and $ba \{a, b, c\}^* b$. This language is ITSL. The context set is $C := \{\langle \sigma, \ltimes, \varepsilon \rangle, \langle \sigma, \varepsilon, \ltimes \rangle, \langle \sigma, \ltimes \sigma', \varepsilon \rangle \mid \sigma, \sigma' \in a, b, c\}$, and the only allowed k-factors are $\ltimes aba \ltimes$ and $\ltimes bab \ltimes$. Now consider the string $s_1 := abc^k bc^k b$, which is not in the concatenation closure of L. Nor is its iteration s_1^m. But the concatenation closure of L does contain $s_2 := s_1^m abs_1^m$, as this is an instance of $ab \{a, b, c\}^* a$ concatenated with $ba \{a, b, c\}^* b$. Every k-context of s_1^m is also a k-context of s_2. Hence every m-factor of s_1^m is also an m-factor of s_2. Therefore it is impossible for any k-ITSL$_m$ grammar G to contain s_2 to the exclusion of s_1. It follows that the concatenation closure of L is not k-ITSL for any k.

Lemma 3. *ITSL is not closed under union.*

Proof. Let $C := \{\langle a, \varepsilon, \varepsilon \rangle, \langle b, \varepsilon, \varepsilon \rangle\}$ and consider the SL$_2$ languages $a^+ b^+$ and $b^+ a^+$. Let L_{ab} and L_{ba} be the respective images of these languages under π_C^{-1} given alphabet $\{a, b, c\}$. That is to say, $L_{ab} := (c^* a)^+ (c^* b)^+ c^*$ and $L_{ba} := (c^* b)^+ (c^* a)^+ c^*$. By definition, L_{ab} and L_{ba} are ITSL languages, but their union L is not. Note that $s_1 := (c^k a)^m c^k \notin L$, whereas $s_2 := s_1^m (c^k b^k)^m c^k \in L$ and $s_3 := (c^k b^k)^m s_1^m c^k \in L$. Every k-context of s_1 also occurs in s_2 and s_3. This implies that no matter what k-context set one picks, all the m-factors of the tier of s_1 are also m-factors of the tiers of s_2 or s_3. As with concatenation closure, this makes it impossible to ban s_1 while allowing for s_2 and s_3.

The same string embedding strategy can also be used for relative complement.

Lemma 4. *ITSL is not closed under relative complement.*

Proof. For simplicity, we only prove non-closure under complement relative to Σ^* (this suffices because Σ^* is ITSL). Let C be as before, and consider the SL_2 language a^+b. The image under π_C^{-1} is the ITSL language $L := (c^*a)^+c^*bc^*$. Consider the string $s_1 := (c^ka)^mc^kbc^k \in L$. The complement $\overline{L}$ of L does not contain s_1, but it contains its mirror immage $s_{-1} := c^kbc^k(ac^k)^m$ and the concatenation of s_1 with itself: $s_{11} := (c^ka)^mc^kbc^k(c^ka)^mc^kbc^k \in \overline{L}$. But as before, every conceivable k-context of s_1 is also a k-context of and s_{-1} and s_{11}. Any illicit m-factor in the tier of s_1 will also occur in the tier of s_{-1} or s_{11}. Again once cannot rule out s_1 without also ruling out s_{-1} or s_{11}, which proves that $\overline{L}$ is not ITSL.

For non-closure under relabeling, a much simpler strategy suffices. Simply consider the SL (and thus ITSL) language $L_{ab} = (ab)^+$. A relabeling that replaces b by a maps L_{ab} to $L_{aa} = (aa)^+$, which isn't even star-free.

Theorem 6. *ITSL is not closed under intersection, union, relative complement, concatenation, and relabelings.*

While these closure properties may seem unappealing from a mathematical perspective, they mirror exactly the closure properties of TSL. This confirms our original claim that ITSL is a natural generalization of TSL. In addition, the lack of most of the canonical closure properties is welcome from a linguistic perspective because natural languages do not seem to display these closure properties either. That said, closure under intersection is a linguistically important property, which is why we explore it in depth in the next section.

5   Intersection Closure of TSL and ITSL

Lack of closure under intersection is problematic as it entails that the complexity of phonological dependencies is no longer constant under factorization. Depending on whether one treats a constraint as a single phenomenon or the interaction of multiple phenomena, the upper bound for phonological complexity will shift. Neither TSL nor ITSL are closed under intersection, yet they both are reasonable formal approximations of phonological dependencies. In order to understand what (I) TSL claims about individual phenomena imply about the complexity of phonology as a whole, we need a good formal understanding of the intersection closure of TSL (Sect. 5.1) and ITSL (Sect. 5.2).

5.1   Intersection Closure of TSL Languages

The intersection of two TSL languages can be regarded as a language that is produced by a single TSL grammar that projects multiple tiers. For this reason,

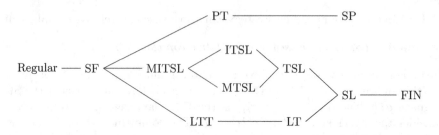

Fig. 1. Proper inclusion relationships of subregular classes. Subsumption goes left-to-right. We establish MTSL, ITSL, and MITSL.

we refer to the intersection closure of TSL as multi-TSL (MTSL). We write n-MTSL$_k$ to indicate a grammar where n is the number of tiers and k is the locality of the tier-constraints. Note that we frequently omit k and n to reduce clutter.

Definition 10. *An n-tier strictly k-local (n-MTSL$_k$) language L is the intersection of n distinct k-TSL languages ($k, n \in \mathbb{N}$).*

MTSL is a proper superclass of TSL, which is witnessed by the language we used to prove non-closure under intersection for ITSL. This also shows that MTSL is not subsumed by ITSL. The opposite does not hold either.

Lemma 5. *ITSL $\not\subseteq$ MTSL.*

Proof. Assume $\Sigma = \{a, b\}$, and consider the language $L_{FL} = a\{a, b\}^*b \cup b\{a, b\}^*a$. This language is ITSL. Suppose L_{FL} were the intersection of n distinct TSL languages $L_1, \ldots, L_n$. Since $a\{a, b\}^*a \notin L$, there would have to be at least one L_i projecting every a on the tier, and banning aa. But then this language also incorrectly rules out aa^+b. Thus, $L \notin n$-MTSL for any number of intersecting TSL languages.

Theorem 7. *MTSL and ITSL are incomparable.*

Regarding the place of MTSL with respect to the other subregular classes, we can reuse most of the previous results. That MTSL $\not\subseteq$ LTT, PT is entailed by TSL $\not\subseteq$ LTT, PT. To see why LTT $\not\subseteq$ MTSL, consider $\Sigma = \{a, b, c\}$ and a sentential logic formula $\varphi : aa \rightarrow bb$ s.t. $L = \{w \in \Sigma^* \mid w \vDash \varphi\}$. Following the same reasoning as in the proof for Theorem 3, it is easy to see that this language is 2-LT (thus, LTT) but not MTSL$_n$. For PT $\not\subseteq$ MTSL, we take the same example and assume that the predicates in φ are based on *precedence* instead of *immediate precedence*. Again following the reasoning in Theorem 3, this language is PT, but not n-MTSL for any n. Finally, MTSL $\subsetneq$ SF follows trivially from the fact that every TSL language is SF [17] and that SF languages are closed under finite intersection.

Theorem 8. *MTSL is incomparable to LT and PT, and MTSL $\subsetneq$ SF.*

5.2   Intersection Closure of ITSL Languages

The definition of MTSL extends in the expected manner to ITSL.

Definition 11 *(MITSL). A multiple m-input local TSL $((m,n)$-MITSL$_k$) language is the intersection of n distinct m-ITSL$_k$ languages $(k, m, n \in \mathbb{N})$.*

Since ITSL is not closed under intersection, we have ITSL $\subsetneq$ MITSL, which in turn implies MTSL $\subsetneq$ MITSL because MTSL and ITSL are incomparable. Just like TSL, MTSL, and ITSL, MITSL is incomparable to LTT and PT. That MITSL $\not\subseteq$ LTT, PT follows from their incomparability to TSL, ITSL and MTSL, which MITSL properly subsumes. For the other direction, we can simply refer to the counter-examples used in Theorem 7, which are not MITSL irrespective of the number of tiers projected by the grammar.

Theorem 9. *MITSL is incomparable to LTT and PT.*

The incomparability to LTT and PT also entails MITSL $\subsetneq$ SF (MITSL $\subseteq$ SF follows from the FO definability of ITSL and the closure of SF under intersection).

Lemma 6. *ITSL $\subsetneq$ MITSL $\subsetneq$ SF.*

This shows that MTSL, ITSL, and MITSL are all natural generalizations of TSL that preserve the relation to other language classes. This extends even to their closure properties: TSL and ITSL have exactly the same closure properties with respect to intersection, union, complement, concatenation, and relabeling, and the multi-tier variants only gain closure under intersection (the proofs for ITSL carry over with simple modifications). In addition, TSL is the natural special case of MITSL with only one tier and ISL$_1$ tier projection.

From a linguistic perspective, this means that even though TSL is inadequate in multiple respects, the insights it yields are preserved with only minor modifications. TSL is not sufficiently expressive for all phonotactic dependencies, but the move from TSL to ITSL is conceptually natural and does not affect common closure properties. TSL complexity results also do not carry over from individual processes to the whole system, but the extension of TSL to MTSL via multiple tiers is linguistically appealing and once again does not affect closure properties or the relation to other language classes. Quite simply, TSL is but one particular point in a whole region of TSL-like classes, all of which behave very similar with respect to closure properties and their relative place in the subregular hierarchy.

6   Learnability Considerations

In this paper we have explored the effects of generalizing the tier projection function for TSL languages to allow for structure-sensitivity. As long as one limits structure-sensitivity to locally bounded contexts, the shift is very natural and mathematically well-behaved. In particular, ITSL allows for additional expressivity while still excluding many unnatural patterns from the classes LT, LTT, SP, PT, and SF (Fig. 1 on page 12).

But generative capacity is not the only linguistically relevant property of language classes. Learnability is also crucial and has profound implications for natural language acquisition [18]. The extensions we have proposed in this paper do not alter the learnability of TSL in the limit from positive text. While the whole class of TSL is not learnable in this paradigm because it properly subsumes the class FIN of all finite languages, TSL_k for $k \geq 0$ is finite and thus learnable [11]. This finiteness also holds for our extensions of TSL as long as all parameters are bounded.

Theorem 10. *Given fixed k, m, and n, (n, m)-MITSL$_k$ languages are learnable in the limit from positive text.*

This still leaves open, though, whether these languages are efficiently learnable. We expect this to be the case given the existence of efficient learners for ISL and TSL [6,21,22]. Moreover, [25] propose an efficient algorithm for MTSL$_2$ building on the notion of a *2-path* exploited by [21]. In a similar fashion, it should be possible to infer local contexts in the projection of tier-segments.

Conjecture 1. (n, m)-MITSL$_k$ languages are efficiently learnable from a polynomial sample size in polynomial time.

The phonotactic phenomena studied so far suggest tight bounds on m, n, k as relevant to the class of human languages [1,15]. Typological explorations thus offer important insights into human learning abilities [8,30].

7 Conclusions

TSL languages have been proposed as a good computational hypothesis for the complexity of phonotactic patterns. However, their tier projection function is too limited because it is context agnostic. A wide range of empirical phenomena can be captured if one equips TSL with an input-strictly local projection mechanism in the sense of Chandlee [5]. The resulting new class of ITSL has the same closure properties as TSL and extends generative capacity only by a small amount. In particular, ITSL occupies a similar position to TSL in the subregular hierarchy.

This paper has explored but one point in a whole region of TSL-like language classes. For instance, we completely omitted OTSL [23] and IOTSL [15]. We also limited ourselves to comparisons to well-established classes such as LTT, ignoring more recently defined classes [13,34]. One major reason for this limit in scope is the lack of fertile characterizations of TSL and ITSL languages. Whereas suffix substitution closure makes it very easy to show that a string language is not strictly local, TSL and ITSL introduce the additional parameter of tiers and contexts that are hard to quantify over in practice. We were able to use string embeddings to create subsumption relations between the contexts and k-factors of specific strings, but this technique is not nearly as versatile as suffix substitution closure. The lack of an equally elegant characterization of TSL and its variants is a serious impediment to a full exploration of the TSL region.

Acknowledgments. This material is based upon work supported by the National Science Foundation under Grant No. BCS-1845344.

References

1. Aksenova, A., Deshmukh, S.: Formal restrictions on multiple tiers. Proc. Soc. Comput. Linguist. (SCiL) **20**(18), 64–73 (2018)
2. Applegate, R.: Ineseno chumash grammar. Ph.D. thesis, UC Berkeley (1972)
3. Baek, H.: Computational representation of unbounded stress patterns: tiers with structural features. In: Proceedings of the 53rd Meeting of the Chicago Linguistic Society (CLS53) (2017)
4. Brzozowski, J.A., Knast, R.: The dot-depth hierarchy of star-free languages is infinite. J. Comput. Syst. Sci. **16**(1), 37–55 (1978)
5. Chandlee, J.: Strictly local phonological processes. Ph.D. thesis, University of Delaware (2014)
6. Chandlee, J., Eyraud, R., Heinz, J.: Learning strictly local subsequential functions. Trans. ACL **2**, 491–503 (2014)
7. Chandlee, J., Heinz, J.: Strict locality and phonological maps. Linguist. Inq. **49**, 23–60 (2018)
8. De Santo, A.: Commentary: developmental constraints on learning artificial grammars with fixed, flexible, and free word order. Front. Psychol. **9**, 276 (2018)
9. Eilenberg, S.: Automata, Languages, and Machines. Academic Press, Inc., Cambridge (1974)
10. Fu, J., Heinz, J., Tanner, H.G.: An algebraic characterization of strictly piecewise languages. In: Ogihara, M., Tarui, J. (eds.) TAMC 2011. LNCS, vol. 6648, pp. 252–263. Springer, Heidelberg (2011). https://doi.org/10.1007/978-3-642-20877-5_26
11. Gold, E.M.: Language identification in the limit. Inf. Control **10**(5), 447–474 (1967)
12. Goldsmith, J.: Autosegmental phonology. Ph.D. thesis, MIT, Cambridge (1976)
13. Graf, T.: The power of locality domains in phonology. Phonology **34**(2), pp. 385–405 (2017). https://doi.org/10.1017/S0052675717000197
14. Graf, T.: Locality domains and phonological c-command over strings. In: 2017 Proceedings of NELS (2018). http://ling.auf.net/lingbuzz/004080
15. Graf, T., Mayer, C.: Sanskrit n-retroflexion is input-output tier-based strictly local. In: 2018 Proceedings of SIGMORPHON (2018)
16. Heinz, J.: The computational nature of phonological generalizations. In: Hyman, L., Plank, F. (eds.) Phonological Typology, chap. 5, pp. 126–195. Phonetics and Phonology, Mouton De Gruyter (2018)
17. Heinz, J., Rawal, C., Tanner, H.: Tier-based strictly local constraints for phonology. In: Proceedings of the ACL 49th: Human Language Technologies: Short Papers - vol. 2, pp. 58–64 (2011). http://dl.acm.org/citation.cfm?id=2002736.2002750
18. Heinz, J., Riggle, J.: Learnability. In: van Oostendorp, M., Ewen, C., Hume, B., Rice, K. (eds.) Blackwell Companion to Phonology. Wiley-Blackwell, Hoboken (2011)

19. Jäger, G., Rogers, J.: Formal language theory: refining the chomsky hierarchy. Philos. Trans. R. Soc. B: Biol. Sci. **367**(1598), 1956–1970 (2012)
20. Jardine, A.: Computationally, tone is different. Phonology (2016). http://udel.edu/~ajardine/files/jardinemscomputationallytoneisdifferent.pdf
21. Jardine, A., Heinz, J.: Learning tier-based strictly 2-local languages. Trans. ACL **4**, 87–98 (2016). https://aclweb.org/anthology/Q/Q16/Q16-1007.pdf
22. Jardine, A., McMullin, K.: Efficient learning of tier-based strictly k-local languages. In: Drewes, F., Martín-Vide, C., Truthe, B. (eds.) LATA 2017. LNCS, vol. 10168, pp. 64–76. Springer, Cham (2017). https://doi.org/10.1007/978-3-319-53733-7_4
23. Mayer, C., Major, T.: A challenge for tier-based strict locality from Uyghur backness harmony. In: Foret, A., Kobele, G., Pogodalla, S. (eds.) FG 2018. LNCS, vol. 10950, pp. 62–83. Springer, Heidelberg (2018). https://doi.org/10.1007/978-3-662-57784-4_4
24. McMullin, K.: Tier-based locality in long-distance phonotactics?: learnability and typology. Ph.D. thesis, University of British Columbia, February (2016). https://doi.org/10.14288/1.0228114
25. McMullin, K., Aksënova, A., De Santo, A. (2019): Learning phonotactic restrictions on multiple tiers. Proc. SCiL **2**(1), pp. 377–378 (2019). https://doi.org/10.7275/s8ym-bx57
26. McNaughton, R., Papert, S.: Counter-Free Automata. MIT Press, Cambridge (1971)
27. Pin, J.E.: Varieties of Formal Languages. Plenum Publishing Co., New York (1986)
28. Rogers, J., et al.: On languages piecewise testable in the strict sense. In: Ebert, C., Jäger, G., Michaelis, J. (eds.) MOL 2007/2009. LNCS (LNAI), vol. 6149, pp. 255–265. Springer, Heidelberg (2010). https://doi.org/10.1007/978-3-642-14322-9_19
29. Rogers, J., Heinz, J., Fero, M., Hurst, J., Lambert, D., Wibel, S.: Cognitive and subregular complexity. In: Morrill, G., Nederhof, M.-J. (eds.) FG 2012-2013. LNCS, vol. 8036, pp. 90–108. Springer, Heidelberg (2013). https://doi.org/10.1007/978-3-642-39998-5_6
30. Rogers, J., Pullum, G.K.: Aural pattern recognition experiments and the subregular hierarchy. J. Logic Lang. Inf. **20**(3), 329–342 (2011)
31. Simon, I.: Piecewise testable events. In: Brakhage, H. (ed.) GI-Fachtagung 1975. LNCS, vol. 33, pp. 214–222. Springer, Heidelberg (1975). https://doi.org/10.1007/3-540-07407-4_23
32. Thomas, W.: Languages, automata, and logic. In: Rozenberg, G., Salomaa, A. (eds.) Handbook of Formal Languages, pp. 389–455. Springer, Heidelberg (1997). https://doi.org/10.1007/978-3-642-59126-6_7
33. Walker, R.: Yaka nasal harmony: spreading or segmental correspondence? Annu. Meet. Berkeley Linguist. Soc. **26**(1), 321–332 (2000). https://doi.org/10.3765/bls.v26i1.1164
34. Yli-Jyrä, A.: Contributions to the theory of finite-state based linguistic grammars. Ph.D. thesis, University of Helsinki (2005). http://www.ling.helsinki.fi/~aylijyra/dissertation/contribu.pdf

Quantificational Subordination
as Anaphora to a Function

Matthew Gotham[(⊠)] [iD]

University of Oxford, Oxford, UK
matthew.gotham@ling-phil.ox.ac.uk

Abstract. In [11], a semantics for cross-sentential and donkey anaphora
is presented that is inspired by approaches using dependent types but
couched in simple type theory with parametric polymorphism. In this
paper, the approach is extended to cover quantificational subordination.
It is argued that the approach enjoys advantages over existing accounts
in type-theoretical semantics.

Keywords: Quantificational subordination · Telescoping · Anaphora ·
Polymorphism

1 Introduction

The history of dynamic semantic theories can be seen as a series of generalizations of what sentence meaning is taken to be, in order to account for the range
of constructions out of which and into which it turns out that a pronoun can
be bound. The pioneering work of [12,14,17] was developed in order to account
for cases like (1). Covariation between donkeys and the interpretation if *it* cannot straightforwardly[1] be accounted for in a static semantic system according
to which the interpretation of a sentence is its truth conditions, or its set of
verifying assignments, as in classical logic.

(1) Every farmer who owns a donkey beats it. [10]

However, via a generalization to the interpretation of a sentence in terms of
a relation between assignments, this covariation can be accounted for. By way
of example, the interpretation of (1) in the system described by [12], assuming
the translation into logical formalism shown in (2), is shown in (3).

(2) $\forall x((\mathsf{farmer}(x) \land \exists y(\mathsf{donkey}(y) \land \mathsf{own}(x,y))) \to \mathsf{beat}(x,y))$

---

[1] 'Straightforwardly', here, means by treating pronouns basically as variables. Static
semantic systems can account for cases like this by treating pronouns as something
else, for example concealed descriptions [7,9].

© Springer-Verlag GmbH Germany, part of Springer Nature 2019
R. Bernardi et al. (Eds.): FG 2019, LNCS 11668, pp. 51–66, 2019.
https://doi.org/10.1007/978-3-662-59648-7_4

$$(3) \quad \left\{ \langle g, g \rangle \mid \begin{cases} h \mid g \approx_{x,y} h \And h(x) \in [\![farmer]\!] \And h(y) \in [\![donkey]\!] \\ \qquad \And \langle h(x), h(y) \rangle \in [\![own]\!] \\ \subseteq \{h \mid \langle h(x), h(y) \rangle \in [\![beat]\!]\} \end{cases} \right\}$$

In turn, however, the first generation of dynamic semantic theories are challenged by examples like (4)–(6). This has led to a further generalization of theories such that now, sentential semantic values are taken to be relations between sets of assignments [4,5] or something more complex altogether [19].

(4) If you give every child a present, some child will open it.   [25]
(5) Every woman bought a book. Most of them read it immediately.   [19]
(6) Every player chooses a pawn. He puts it on square one.   [12]

What examples like these have in common is that the anaphoric relationships that we have to capture seem to depend on functional relationships established in the first clause: between children and presents they are given in (4), between women and books they bought in (5), and between players and pieces they chose in (6). So, second-generation dynamic semantic theories look for ways of preserving those relationships, either by making any output set of assignments for the first clause such that it guarantees that the relationships are preserved [4,5], or by allowing the update that the first sentence engenders to be stored so that it can be reintroduced at the appropriate point in the second sentence [19].

There is an alternative approach, though, which is to see the semantic values of sentences as producing the very functions themselves, not mediated via variable assignments. This is the approach taken in type-theoretical semantics (TTS).

2   Type-Theoretical Semantics

Type-theoretical semantics is a proof-theoretic variety of logical semantics in which the language of types, rather than terms, is the meaning representation language. That is to say, instead of the meaning of a sentence (say) being represented by a term (of a type), which in turn is interpreted in a model, in TTS (in the simplest case) the meaning of a sentence is a type T, where the inhabitants of T are the (intuitionistic) proofs of T. Building on previous work by [29] showing the application to sentences like (1), [25] used the intuitionistic type theory (ITT) of [23] to give a detailed analysis of many natural language phenomena, which has been followed up in recent years by several authors, summarized in [6]. Formation, introduction and elimination rules for the crucial ITT type constructors Σ and Π are shown in Fig. 1. Σ is a generalization of $\times$, from pairs to dependent pairs, and Π is a generalization of $\rightarrow$, from functions to dependent functions.

$$\Sigma \qquad\qquad\qquad \Pi$$

$$[x : A]^n \qquad\qquad\qquad [x : A]^n$$
$$\vdots \qquad\qquad\qquad\qquad \vdots$$

Formation
$$\dfrac{x : A \quad B : \textbf{type}}{(\Sigma x : A)B : \textbf{type}}\ n \qquad\qquad \dfrac{x : A \quad B : \textbf{type}}{(\Pi x : A)B : \textbf{type}}\ n$$

$$[x : A]^n$$
$$\vdots$$

Introduction
$$\dfrac{a : A \quad b : B[a/x]}{(a,b) : (\Sigma x : A)B} \qquad\qquad \dfrac{b : B}{\lambda x.b : (\Pi x : A)B}\ n$$

Elimination
$$\dfrac{u : (\Sigma x : A)B}{\pi_1(u) : A} \quad \dfrac{u : (\Sigma x : A)B}{\pi_2(u) : B[\pi_1(u)/x]} \qquad \dfrac{a : A \quad f : (\Pi x : A)B}{f(a) : B[a/x]}$$

Fig. 1. Formation, introduction and elimination rules for Σ and Π types

A possible TTS representation for (1) is shown in (7).[2] The sentence is taken to be true if and only if there is something of the type shown (the type is inhabited), i.e. a function mapping every tuple $\langle a, \langle b, \langle\langle c, d\rangle, e\rangle\rangle\rangle$, where b is a proof that a is a farmer, d is a proof that c is a donkey, and e is a proof that a owns c, to a proof that a beats c.

(7) $\quad (\Pi u : (\Sigma x : e)(\text{FARMER}(x) \times (\Sigma v : (\Sigma z : e)\,\text{DONKEY}(z))\,\text{OWN}(x, \pi_1(v))))$
$\qquad \text{BEAT}(\pi_1(u), \pi_1(\pi_1(\pi_2(\pi_2(u)))))$

As ([25], Sect. 3.7) points out, the fact that in TTS a universal statement expresses a function means that the functional dependency in cases like (4) can easily be expressed, as shown in (8).

(8) $\quad (\Pi f : (\Pi u : (\Sigma x : e)\,\text{CHILD}(x))$
$\qquad\quad (\Sigma v : (\Sigma y : e)\,\text{PRESENT}(y))\,\text{GIVE}((you, \pi_1(v), \pi_1(u)))$
$\qquad\quad (\Sigma w : (\Sigma z : e)\,\text{CHILD}(z))\,\text{OPEN}(\pi_1(w), \pi_1(\pi_1(f(w))))$

Given any function f mapping every pair $\langle a, b \rangle$, where b is a proof that a is a child, to a tuple $\langle\langle c, d\rangle, e\rangle$, where d is a proof that c is a present and e is a proof that you give c to a, (8) expresses the existence of a function mapping f to a tuple $\langle\langle g, h\rangle, i\rangle$ where h is a proof that g is a child and i is a proof that g opens $\pi_1(\pi_1(f(g, h)))$. So TTS automatically has the ability to capture antecedent-pronoun relationships that first-generation dynamic semantic theories struggle with. This point has been futher explored in [31,32].

Nevertheless, the direct applicability of this antecedence-to-a-function strategy is limited. In the case of (4), it was actually crucial that an appropriate

---

[2] Following the approach of [2] rather than that of [29] or [25], i.e. treating the interpretation of e.g. *donkey* not as the type DONKEY (inhabited by donkeys), but rather as the dependent type DONKEY(x) (for any $x : e$), inhabited by proofs that x is a donkey.

argument for the function (*some child*) was present in the second sentence. In a discussion of (6), ([25], p. 73) acknowledges as much in saying that 'the only way to interpret the text [...] is by treating the pronoun *he* as an abbreviation of *every player*'. Obviously, this 'abbreviation' strategy is unsatisfactory. Below, I will show how it can be improved by taking a different tack.

3   The Proposal

In [11], an implementation of the ideas behind TTS is given in simple type theory with polymorphism, once again treating the language of terms as the meaning representation language. Taking the perspective outlined in [11], equivalents for (7) and (8) are shown in (9) and (10), respectively. The reader can verify that these represent the same meanings as given in (7) and (8) according to the glosses provided, on the understanding that, following a suggestion by ([25], Sect. 2.26), in turn picked up by [8], we allow eventualities (type v) to serve as 'proof objects'.[3] N.B. in what follows, in order to save space, application will be written as fa rather than $f(a)$ and left and right projections will be written as a_0 and a_1 rather than $\pi_1(a)$ and $\pi_2(a)$, respectively. In type annotations, $\times$ binds more tightly than $\rightarrow$, and both associate to the right.

(9) $\exists f^{e\times e\times v\rightarrow v}.\forall u^{e\times e\times v}.\mathsf{farmer}\,u_0 \wedge \mathsf{donkey}(u_1)_0 \wedge \mathsf{own}\,u \rightarrow \mathsf{beat}(u_0, (u_1)_0, fu)$

(10) $\exists f^{\tau\rightarrow e\times v}.\forall g^{\tau}.(\forall x^e.\mathsf{child}\,x \rightarrow (\mathsf{present}(gx)_0 \wedge \mathsf{give}(\mathsf{you}, (gx)_0, x, (gx)_1)))$
$\rightarrow (\mathsf{child}(fg)_0 \wedge \mathsf{open}((fg)_0, (g(fg)_0)_0, (fg)_1))$

$$\text{where } \tau := e \rightarrow e \times v$$

Like TTS, the system described in [11] does not straightforwardly have the means to account for the examples (5)–(6). In this section an extension is described that does so, on the basis of the lexicon shown in Fig. 2.

3.1   Syntax and Semantics

The syntactic theory assumed here is a modified version of Combinatory Categorial Grammar (CCG, [27]) according to which syntactic categories are potentially parameterized by types. Lexical entries are pairs $M : C$ of meaning M and category C such that the type of M is $\mathrm{Ty}(C)$, where Ty is as defined in (11). The combinatory rules for the fragment are shown in Fig. 3. Note that we have adopted the G rule from [15] for passing pronominal dependencies (without adopting the theory of pronouns described in [15]), to which we have added a rule X of exchange.[4]

---

[3] I only claim that eventualities (states or Davidsonian events) can be operationalized in this way, not that this is the only way to make sense of proof objects in a model-theoretic perspective. Other options worth considering would be situations [21] or even, as one reviewer suggests, to fill the proof object slot with a dummy object of the unit type (as I have done for common nouns).

[4] In the G and X rules, |could be / or \.

$$if \rightsquigarrow \lambda p^{\alpha \rightarrow \beta \rightarrow t}.\lambda q^{\alpha \times \beta \rightarrow \gamma \rightarrow t}.\lambda i^{\alpha}.\lambda f^{\beta \rightarrow \gamma}.\mathsf{dom} f = (\lambda b^{\beta}.pib)$$
$$\wedge\, \forall b^{\beta}.\mathsf{dom} fb \rightarrow q(i,b)(fb)$$
$$: (\mathsf{S}_{\alpha,\beta \rightarrow \gamma}/\mathsf{S}_{\alpha \times \beta,\gamma})/\mathsf{S}_{\alpha,\beta}$$

$$and,; \rightsquigarrow \lambda p^{\alpha \rightarrow \beta \rightarrow t}.\lambda q^{\alpha \times \beta \rightarrow \gamma \rightarrow t}.\lambda i^{\alpha}.\lambda o^{\beta \times \gamma}.pio_0 \wedge q(i,o_0)o_1$$
$$: (\mathsf{S}_{\alpha,\beta \times \gamma}/\mathsf{S}_{\alpha \times \beta,\gamma})\backslash \mathsf{S}_{\alpha,\beta}$$

$$a \rightsquigarrow \lambda P^{\alpha \rightarrow e \times \beta \rightarrow t}.\lambda V^{e \rightarrow \alpha \times e \times \beta \rightarrow \gamma \rightarrow t}.\lambda i^{\alpha}.\lambda u^{(e \times \beta) \times \gamma}.Piu_0 \wedge V(u_0)_0(i,u_0)u_1$$
$$: (\mathsf{S}_{\alpha,(e \times \beta) \times \gamma}/(\mathsf{S}_{\alpha \times e \times \beta,\gamma}\backslash \mathrm{NP}))/\mathrm{N}_{\alpha,e \times \beta}$$

$$child \rightsquigarrow \lambda i^{\alpha}.\lambda u^{e \times 1}.\mathsf{child}\, u_0 : \mathrm{N}_{\alpha,e \times 1}$$

$$bought \rightsquigarrow \lambda D^{\tau'' \rightarrow \beta \rightarrow \gamma \rightarrow t}.\lambda x^e.D(\lambda y^e.\lambda i^{\alpha}.\lambda e^v.\mathsf{buy}(x,y,e))$$
$$: (\mathsf{S}_{\beta,\gamma}\backslash \mathrm{NP})/(\mathsf{S}_{\beta,\gamma}/(\mathsf{S}_{\alpha,v}\backslash \mathrm{NP}))$$

$$who \rightsquigarrow \lambda V^{\tau'}.\lambda P^{\tau}.\lambda i^{\alpha}.\lambda o^{e \times \beta \times \gamma}.Pi(o_0,(o_1)_0) \wedge Vo_0(i,(o_1)_0)(o_1)_1$$
$$: (\mathrm{N}_{\alpha,e \times \beta \times \gamma}\backslash \mathrm{N}_{\alpha,e \times \beta})/(\mathsf{S}_{\alpha \times e \times \beta,\gamma}\backslash \mathrm{NP})$$

$$he,it \rightsquigarrow \lambda g^{\alpha \rightarrow e}.\lambda V^{e \rightarrow \alpha \rightarrow \beta \rightarrow t}.\lambda i^{\alpha}.V(gi)i : (\mathsf{S}_{\alpha,\beta}/(\mathsf{S}_{\alpha,\beta}\backslash \mathrm{NP}))^{\mathrm{NP}\alpha}$$

$$they \rightsquigarrow \lambda G^{\alpha \rightarrow e \rightarrow t}.\lambda V^{e \rightarrow \alpha \rightarrow \beta \rightarrow t}.\lambda i^{\alpha}.\lambda f^{e \rightarrow \beta}.\mathsf{dom} f = (\lambda y^e.Giy)$$
$$\wedge\, \forall x^e.\mathsf{dom} fx \rightarrow Vxi(fx)$$
$$: (\mathsf{S}_{\alpha,e \rightarrow \beta}/(\mathsf{S}_{\alpha,\beta}\backslash \mathrm{NP}))^{\mathrm{NPL}\alpha}$$

$$of\ them \rightsquigarrow \lambda G^{\alpha \rightarrow e \rightarrow t}.\lambda i^{\alpha}.\lambda u^{e \times 1}.Giu_0 : (\mathrm{N}_{\alpha,e \times 1})^{\mathrm{NPL}\alpha}$$

$$det_{\mathrm{weak}} \rightsquigarrow \lambda P^{\tau}.\lambda V^{\tau'}.\lambda i^{\alpha}.\lambda f^{e \times \beta \rightarrow \gamma}.\mathsf{dom} f \subseteq (\lambda v^{e \times \beta}.Piv)$$
$$\wedge\, \mathsf{det}(\lambda x^e.\exists b^{\beta}.Pi(x,b))(\lambda x^e.\exists b^{\beta}.\mathsf{dom} f(x,b))$$
$$\wedge\, (\forall a^{e \times \beta}.\mathsf{dom} fa \rightarrow Va_0(i,a)(fa))$$
$$\wedge\, (\neg \exists Y^{e \times \beta \rightarrow t}.\mathsf{dom} f \subsetneq Y \wedge Y \subseteq (\lambda v^{e \times \beta}.Piv)$$
$$\wedge\, \forall a^{e \times \beta}.Ya \rightarrow \exists c^{\gamma}.Va_0(i,a)c)$$

$$det_{\mathrm{strong}} \rightsquigarrow \lambda P^{\tau}.\lambda V^{\tau'}.\lambda i^{\alpha}.\lambda f^{e \times \beta \rightarrow \gamma}.\mathsf{dom} f \subseteq (\lambda v^{e \times \beta}.Piv)$$
$$\wedge\, \mathsf{det}(\lambda x^e.\exists b^{\beta}.Pi(x,b))(\lambda x^e.\exists b^{\beta}.\mathsf{dom} f(x,b))$$
$$\wedge\, (\forall a^{e \times \beta}.\mathsf{dom} fa \rightarrow Va_0(i,a)(fa))$$
$$\wedge\, (\forall x^e.\forall b^{\beta}.(Pi(x,b) \wedge \exists c^{\beta}.\mathsf{dom} f(x,c))$$
$$\rightarrow \mathsf{dom} f(x,b))$$
$$\wedge\, (\neg \exists Y^{e \times \beta \rightarrow t}.\mathsf{dom} f \subsetneq Y \wedge Y \subseteq (\lambda v^{e \times \beta}.Piv)$$
$$\wedge\, \forall a^{e \times \beta}.Ya \rightarrow \exists c^{\gamma}.Va_0(i,a)c)$$
$$: (\mathsf{S}_{\alpha,e \times \beta \rightarrow \gamma}/(\mathsf{S}_{\alpha \times e \times \beta,\gamma}\backslash \mathrm{NP}))/\mathrm{N}_{\alpha,e \times \beta}$$

$$[\mathrm{close}] := \lambda p^{1 \rightarrow \alpha \rightarrow t}.\exists a^{\alpha}.p * a : \mathsf{S}/\mathsf{S}_{1,\alpha}$$

where $\tau := \alpha \rightarrow e \times \beta \rightarrow t$, $\tau' := e \rightarrow \alpha \times e \times \beta \rightarrow \gamma \rightarrow t$ and $\tau'' := e \rightarrow \alpha \rightarrow v \rightarrow t$

Fig. 2. The lexicon

(11) For any types α, β and any categories A, B:
- $\mathrm{Ty}(\mathsf{S}_{\alpha,\beta}) = \mathrm{Ty}(\mathrm{N}_{\alpha,\beta}) = \alpha \rightarrow \beta \rightarrow t$
- $\mathrm{Ty}(\mathsf{S}) = t$
- $\mathrm{Ty}(\mathrm{NP}_{\alpha}) = \alpha \rightarrow e$
- $\mathrm{Ty}(\mathrm{NP}) = e$
- $\mathrm{Ty}(\mathrm{NPL}_{\alpha}) = \alpha \rightarrow e \rightarrow t$
- $\mathrm{Ty}(\mathrm{NPL}) = e \rightarrow t$
- $\mathrm{Ty}(A\backslash B) = \mathrm{Ty}(A/B) = \mathrm{Ty}(A^B) = \mathrm{Ty}(B) \rightarrow \mathrm{Ty}(A)$

$$\frac{f : B/A \quad a : A}{fa : B} > \qquad \frac{f : A \mid B}{\lambda g^{\mathrm{Ty}(C) \to \mathrm{Ty}(B)}.\lambda c^{\mathrm{Ty}(C)}.f(gc) : A^C \mid B^C} \; G$$

$$\frac{a : A \quad f : B\backslash A}{fa : B} < \qquad \frac{f : (A \mid B)^C}{\lambda b^{\mathrm{Ty}(B)}.\lambda c^{\mathrm{Ty}(C)}.fcb : A^C \mid B} \; X$$

Fig. 3. CCG rules used

On the semantic side, lexical entries are also parameterized by type, so we can see Fig. 2 as in effect giving us schemata over lexical entries. The base types are e (entities), v (eventualities), 1 (unit) and t (booleans), and the type (meta)variables range over the closure of this set under the type constructors $\times$ and $\to$. Furthermore, we are assuming a partial theory of types, as described for example in [13, Sect. 4]. For each base type τ the undefined object of type τ, $\star^\tau$, is stipulated,[5] and then for complex types undefined objects are as specified in (12).[6]

(12) For any types α, β:
  - $\star^{\alpha \times \beta} := \left(\star^\alpha, \star^\beta\right)$
  - $\star^{\alpha \to \beta} :=$ the unique $f :: \alpha \to \beta$ such that for all $a :: \alpha$, $fa = \star^\beta$

This notion of definedness then features in the definition of dom, shown in (13) and used in the analysis of quantificational subordination (dom is mnemonic for 'domain').

(13) For any types α, β and term $f :: \alpha \to \beta$, $\mathrm{dom} f := \lambda a^\alpha.fa \neq \star^\beta$.

Finally, note the general lexical entries for (strong and weak) determiners, which represents an advance on the ad-hoc entries given in [11]. In these entries, det is the meaning of the determiner understood as a relation between sets, in terms of generalized quantifier theory. In those terms, the meanings given are roughly equivalent to saying that a sentence [[D N] VP], assuming that $[\![D]\!]$ is a relation between sets of entities, $[\![N]\!]$ is a set of entities and $[\![VP]\!]$ is a relation between entities and events, expresses the existence of a function f with domain X, where X is a witness set of $[\![D]\!] ([\![N]\!])$,[7] such that f maps every $x \in X$ to some e such that $\langle x, e \rangle \in [\![VP]\!]$, and there is no other set Y such that $X \subsetneq Y \subseteq [\![N]\!]$ and $Y \subseteq \{y \mid \text{there is an } e \text{ such that } \langle y, e \rangle \in [\![VP]\!]\}$. So, for example, *two boys jump* expresses a function f with domain X, where X is a set of two boys, such that f maps every $x \in X$ to an event of x jumping, and there is no proper superset Y of X such that Y is a set of boys and, for every $y \in Y$, there is an event of y jumping.

---

[5] Space precludes proper discussion of various technical questions here; suffice to say that we do have a third truth value but we do not have a second element of the unit type.

[6] $\star$ must not be confused with $*$, which is the unique object of the unit type.

[7] That is, a witness set in the sense of ([1], Sect. 4.9), i.e. a set S such that $S \subseteq [\![N]\!]$ and $\langle [\![N]\!], S \rangle \in [\![D]\!]$.

3.2 Derivations

We now are in a position to show derivations for some examples. First, the simple donkey sentence (1), the derivation of which is given in (14). $GQ_{\alpha,\beta,\gamma,\delta}$ is an abbreviation for $S_{\gamma,\delta}/(S_{\alpha,\beta}\backslash NP)$ and $GQ_{\alpha,\beta*2}$ is an abbreviation for $GQ_{\alpha,\beta,\alpha,\beta}$.

(14) Derivation of (1). Let $\epsilon := e \times 1$, $\sigma := e \times 1 \times \epsilon$ and $\tau := e \times 1 \times e \times 1 \times v$. Then:

$$
\cfrac{
\cfrac{\overset{owns}{(S_{1\times\epsilon,e\times1\times v}\backslash NP)/GQ_{1\times\sigma,v,1\times\epsilon,e\times1\times v}} \quad \cfrac{\overset{a}{GQ_{1\times\sigma,v,1\times\epsilon,e\times1\times v}/N_{1\times\epsilon,\epsilon}} \quad \overset{donkey}{N_{1\times\epsilon,\epsilon}}}{GQ_{1\times\sigma,v,1\times\epsilon,e\times1\times v}}\ {\scriptstyle >}}
{S_{1\times\epsilon,e\times1\times v}\backslash NP}
$$

$$
\cfrac{
\cfrac{\overset{every}{GQ_{1\times\tau,v,1,\tau\to v}/N_{1,\tau}} \quad \cfrac{\overset{farmer}{N_{1,\epsilon}} \quad \cfrac{\overset{who}{(N_{1,\tau}\backslash N_{1,\epsilon})/(S_{1\times\epsilon,e\times1\times v}\backslash NP)} \quad \overset{owns\ a\ donkey}{S_{1\times\epsilon,e\times1\times v}\backslash NP}}{N_{1,\tau}\backslash N_{1,\epsilon}}\ {\scriptstyle <}}{N_{1,\tau}}\ {\scriptstyle >}}
{\cfrac{S_{1,\tau\to v}/(S_{1\times\tau,v}\backslash NP)}{(S_{1,\tau\to v})^{NP1\times\tau}/(S_{1\times\tau,v}\backslash NP)^{NP1\times\tau}}\ G}
$$

$$
\cfrac{\cfrac{\overset{beats}{(S_{1\times\tau,v}\backslash NP)/GQ_{1\times\tau,v*2}}}{(S_{1\times\tau,v}\backslash NP)^{NP1\times\tau}/(GQ_{1\times\tau,v*2})^{NP1\times\tau}}\ G \quad \cfrac{\overset{it}{(GQ_{1\times\tau,v*2})^{NP1\times\tau}}}{}}{(S_{1\times\tau,v}\backslash NP)^{NP1\times\tau}}\ {\scriptstyle >}
$$

$$
\cfrac{
\cfrac{\overset{[close]}{\cfrac{S/(S_{1,\tau\to v})}{S^{NP1\times\tau}/(S_{1,\tau\to v})^{NP1\times\tau}}}\ G \quad \cfrac{\overset{every\ farmer\ who\ owns\ a\ donkey}{(S_{1,\tau\to v})^{NP1\times\tau}/(S_{1\times\tau,v}\backslash NP)^{NP1\times\tau}} \quad \overset{beats\ it}{(S_{1\times\tau,v}\backslash NP)^{NP1\times\tau}}}{(S_{1,\tau\to v})^{NP1\times\tau}}\ {\scriptstyle >}}
{S^{NP1\times\tau}}\ {\scriptstyle >}
$$

Weak interpretation:

$$\lambda g^{1\times\tau\to e}.\exists f^{\tau\to v}.\,\text{dom}\,f \subseteq (\lambda u^\tau.\text{farmer}\,u_0 \wedge \text{donkey}((u_1)_1)_0$$
$$\wedge\,\text{own}(u_0,((u_1)_1)_0,((u_1)_1)_1))$$
$$\wedge\,\text{every}(\lambda x^e.\text{farmer}\,x \wedge \exists z^e.\text{donkey}\,z \wedge \exists e^v.\text{own}(x,z,e))$$
$$(\lambda x^e.\exists o^{1\times e\times1\times v}.\text{dom}\,f(x,o))$$
$$\wedge\,\forall a^\tau.\text{dom}\,fa \to \text{beat}(a_0,(g(*,a)),(fa))$$

Strong interpretation:

$$\lambda g^{1\times\tau\to e}.\exists f^{\tau\to v}.\operatorname{dom} f \subseteq (\lambda u^{\tau}.\mathsf{farmer}\, u_0 \wedge \mathsf{donkey}((u_1)_1)_0$$
$$\wedge\, \mathsf{own}(u_0,((u_1)_1)_0,((u_1)_1)_1))$$
$$\wedge\, \mathsf{every}(\lambda x^e.\mathsf{farmer}\, x \wedge \exists z^e.\mathsf{donkey}\, z \wedge \exists e^v.\mathsf{own}(x,z,e))$$
$$(\lambda x^e.\exists o^{1\times e\times 1\times v}.\operatorname{dom} f(x,o))$$
$$\wedge\, (\forall a^{\tau}.\operatorname{dom} fa \to \mathsf{beat}(a_0,(g(*,a)),(fa)))$$
$$\wedge\, \forall x^e.\forall z^e.\forall e^v.(\mathsf{farmer}\, x \wedge \mathsf{donkey}\, z \wedge \mathsf{own}(x,z,e)$$
$$\wedge\, \exists v^{e\times v}.\operatorname{dom} f(x,v)) \to \operatorname{dom} f(x,z,e)$$

The open abstraction λg represents anaphoric resolution, and so is resolved contextually. In this case, the function that gets the right resolution is shown in (15).

(15) $\lambda i^{1\times e\times 1\times e\times 1\times v}.(((i_1)_1)_1)_0$

The strong interpretation differs from the weak in requiring that, if some farmer x and donkey z that x owns are in the domain of f, then so are x and y for every donkey y that x owns. With the resolution shown in (15) applied, the strong interpretation shown above is equivalent to the one shown in (9). In neither case is the final clause of the definition of a determiner from Fig. 2 shown, because this maximality clause only makes a truth-conditional difference for non-monotone-increasing determiners.

We will come back to the issue of appropriate anaphoric resolution after considering our next example, (5), in (16).

(16) Derivation of (5). Let $\epsilon := e \times 1, \sigma := \epsilon \to \epsilon \times v$ and $\tau := 1 \times \sigma \times \epsilon$. Then:

$$\cfrac{\cfrac{\textit{bought}}{(\mathrm{S}_{1\times\epsilon,\epsilon\times v}\mathrm{NP})/\mathrm{GQ}_{(1\times\epsilon)\times\epsilon,v,1\times\epsilon,\epsilon\times v}} \quad \cfrac{\cfrac{\textit{a}}{\mathrm{GQ}_{(1\times\epsilon)\times\epsilon,v,1\times\epsilon,\epsilon\times v}/\mathrm{N}_{1\times\epsilon,\epsilon}} \quad \cfrac{\textit{book}}{\mathrm{N}_{1\times\epsilon,\epsilon}}}{\mathrm{GQ}_{(1\times\epsilon)\times\epsilon,v,1\times\epsilon,\epsilon\times v}}>}{\mathrm{S}_{1\times\epsilon,\epsilon\times v}\mathrm{NP}}>$$

$$\cfrac{\cfrac{\cfrac{\textit{every}}{\mathrm{GQ}_{1\times\epsilon,\epsilon\times v,1,\sigma}/\mathrm{N}_{1,\epsilon}} \quad \cfrac{\textit{woman}}{\mathrm{N}_{1,\epsilon}}}{\mathrm{S}_{1,\sigma}/(\mathrm{S}_{1\times\epsilon,\epsilon\times v}\backslash\mathrm{NP})}> \quad \cfrac{\textit{bought a book}}{\mathrm{S}_{1\times\epsilon,\epsilon\times v}\mathrm{NP}}}{\cfrac{\mathrm{S}_{1,\sigma}}{\cfrac{\mathrm{S}_{1,\sigma\times(\epsilon\to v)}/\mathrm{S}_{1\times\sigma,\epsilon\to v}}{\cfrac{(\mathrm{S}_{1,\sigma\times(\epsilon\to v)})^{\mathrm{NPL}_{1\times\sigma}}/(\mathrm{S}_{1\times\sigma,\epsilon\to v})^{\mathrm{NPL}_{1\times\sigma}}}{((\mathrm{S}_{1,\sigma\times(\epsilon\to v)})^{\mathrm{NPL}_{1\times\sigma}})^{\mathrm{NP}_{\tau}}/((\mathrm{S}_{1\times\sigma,\epsilon\to v})^{\mathrm{NPL}_{1\times\sigma}})^{\mathrm{NP}_{\tau}}}\, G}\, G}\quad ; \quad (\mathrm{S}_{1,\sigma\times(\epsilon\to v)}/\mathrm{S}_{1\times\sigma,\epsilon\to v})\backslash\mathrm{S}_{1,\sigma}}<$$

$$\frac{\dfrac{\overset{\textit{read}}{(S_{\tau,v}\backslash NP)/GQ_{\tau,v*2}}}{(S_{\tau,v}\backslash NP)^{NP_\tau}/(GQ_{\tau,v*2})^{NP_\tau}}\ G\quad \overset{\textit{it}}{(GQ_{\tau,v*2})^{NP_\tau}}}{(S_{\tau,v}\backslash NP)^{NP_\tau}}\ >$$

$$\frac{\dfrac{\overset{\textit{most}}{GQ_{\tau,v,1\times\sigma,\epsilon\to v}/N_{1\times\sigma,\epsilon}}}{(GQ_{\tau,v,1\times\sigma,\epsilon\to v})^{NPL1\times\sigma}/(N_{1\times\sigma,\epsilon})^{NPL1\times\sigma}}\ G\quad \overset{\textit{of them}}{(N_{1\times\sigma,\epsilon})^{NPL1\times\sigma}}}{(S_{1\times\sigma,\epsilon\to v}/(S_{\tau,v}\backslash NP))^{NPL1\times\sigma}}\ >$$

$$\frac{\dfrac{(S_{1\times\sigma,\epsilon\to v})^{NPL1\times\sigma}/(S_{\tau,v}\backslash NP)}{((S_{1\times\sigma,\epsilon\to v})^{NPL1\times\sigma})^{NP_\tau}/(S_{\tau,v}\backslash NP)^{NP_\tau}}\ X\quad G\quad \overset{\textit{read it}}{(S_{\tau,v}\backslash NP)^{NP_\tau}}}{((S_{1\times\sigma,\epsilon\to v})^{NPL1\times\sigma})^{NP_\tau}}\ >$$

$$\frac{\overset{\textit{every woman bought a book;}}{((S_{1,\sigma\times(\epsilon\to v)})^{NPL1\times\sigma})^{NP_\tau}/((S_{1\times\sigma,\epsilon\to v})^{NPL1\times\sigma})^{NP_\tau}}\quad \overset{\textit{most of them read it}}{((S_{1\times\sigma,\epsilon\to v})^{NPL1\times\sigma})^{NP_\tau}}}{((S_{1,\sigma\times(\epsilon\to v)})^{NPL1\times\sigma})^{NP_\tau}}\ >$$

$$\vdots\ [close]$$
$$(S^{NPL1\times\sigma})^{NP_\tau}$$

Weak interpretation:

$$\lambda g^{\tau\to e}.\lambda G^{1\times\sigma\to e\to t}.\exists W^{\sigma\times(e\times 1\to v)}.\mathrm{dom}\,W_0 = (\lambda o^{e\times 1}.\mathsf{woman}\,o_0)$$
$$\wedge\ (\forall u^{e\times 1}.\mathrm{dom}\,W_0 u \to (\mathsf{book}((W_0 u)_0)_0 \wedge \mathsf{read}(u_0,((W_0 u)_0)_0,(W_0 u)_1)))$$
$$\wedge\ \mathrm{dom}\,W_1 \subseteq (\lambda v^{e\times 1}.G(*,W_0)v_0)$$
$$\wedge\ \mathsf{most}(\lambda x^e.G(*,W_0)x)(\lambda x^e.\mathrm{dom}\,W_1(x,*))$$
$$\wedge\ \forall a^{e\times 1}.\mathrm{dom}\,W_1 a \to \mathsf{read}(a_0,g(*,W_0,a),(W_1 a))$$

In this case there are two open abstractions to resolve, for *it* and *of them* respectively. The appropriate resolutions are shown in (17) and (18) respectively.

(17) $\lambda n^{1\times(e\times 1\to(e\times 1)\times v)\times e\times 1}.(((n_1)_0\,(n_1)_1)_0)_0$

(18) $\lambda m^{1\times(e\times 1\to(e\times 1)\times v)}.\lambda x^e.\exists i^1.\mathrm{dom}\,m_1(x,i)$
$\equiv \lambda m^{1\times(e\times 1\to(e\times 1)\times v)}.\lambda x^e.\mathrm{dom}\,m_1(x,*)$

With those resolutions in place, the (weak) interpretation of the sentence is:

$$\exists W^{\sigma\times(e\times 1\to v)}.\mathrm{dom}\,W_0 = (\lambda o^{e\times 1}.\mathsf{woman}\,o_0)$$
$$\wedge\ (\forall u^{e\times 1}.\mathrm{dom}\,W_0 u \to (\mathsf{book}((W_0 u)_0)_0 \wedge \mathsf{read}(u_0,((W_0 u)_0)_0,(W_0 u)_1)))$$
$$\wedge\ \mathrm{dom}\,W_1 \subseteq \mathrm{dom}\,W_0 \wedge \mathsf{most}(\lambda x^e.\mathrm{dom}\,W_0(x,*))(\lambda x^e.\mathrm{dom}\,W_1(x,*))$$
$$\wedge\ \forall a^{e\times 1}.\mathrm{dom}\,W_1 a \to \mathsf{read}(a_0,(W_0 a)_0,(W_1 a))$$

This expresses the existence of a pair of functions, the first f mapping every woman to a book she bought and the second mapping most things x in the domain of f (i.e. most women) to an event of x reading fx.

The anaphoric resolution functions (15), (17) and (18) are all natural resolution functions (NRFs) according to the definition shown in (19).

(19) The set of NRFs is the smallest set such that, for any types α, β and γ and any terms $F :: \alpha \to \beta \to \gamma$, $G :: \beta \to \gamma$ and $H :: \alpha \to \beta$:
- $\lambda a^\alpha.a$ is an NRF
- $\lambda A^{\alpha \times \beta}.A_0$ is an NRF
- $\lambda A^{\alpha \times \beta}.A_1$ is an NRF
- $\lambda X^{\alpha \times \beta \to t}.\lambda a^\alpha.\exists b^\beta.X(a, b)$ is an NRF
- $\lambda X^{\alpha \times \beta \to t}.\lambda b^\alpha.\exists a^\beta.X(a, b)$ is an NRF
- $\lambda f^{\alpha \to \beta}.\mathrm{dom}\, f$ is an NRF
- $\lambda a^\alpha.G(Ha)$ is an NRF if G and H are NRFs
- $\lambda a^\alpha.Fa(Ha)$ is an NRF if F and H are NRFs

Of course, this definition has been formulated post-hoc. Nevertheless, there is a naturalness to it: a resolution function can select projections, sets of projections, and the domain of a function, and can apply one thing it selects to another when the types match.

4 Telescoping

To deal with example (6), however (repeated below), additional machinery is required.

(6) Every player selects a pawn. He puts it on square one.

The reason that examples like (6), sometimes called 'telescoping' [26], pose a particular challenge is that there is nothing in the second sentence that is explicitly anaphoric on the function expressed by the first sentence. In contrast, the seemingly equivalent sentence (20) could be dealt with in basically the same way as (5), by making *they* anaphoric to the domain of the function mapping every player to the pawn he selects, i.e. the set of players.

(20) Every player selects a pawn. They put it on square one.

4.1 Covert Subordination

In order to deal with examples like (6), [26] posits the existence of a covert adverbial at the start of the second sentence, meaning something like 'in every case'. We will adopt essentially the same strategy. In (21) we postulate a silent subordinating operator that, when applied to the usual sentential conjunction shown in Fig. 2, gives (22) as an alternative, subordinating, sentential conjunction.

(21) $\lambda C^{(\alpha \to (\beta \to \gamma) \to t) \to (\alpha \times (\beta \to \delta) \to (\beta \to \delta) \to t) \to \alpha \to (\beta \to \gamma) \times (\beta \to \delta) \to t}.$
$\quad \lambda p^{\alpha \to (\beta \to \gamma) \to t}.\lambda q^{\alpha \times \beta \times (\beta \to \gamma) \to \delta \to t}.Cp\big(\lambda o^{\alpha \times (\beta \to \gamma)}.\lambda f^{\beta \to \delta}.\mathrm{dom}\, o_1 = \mathrm{dom}\, f$
$\qquad\qquad\qquad\qquad\qquad \wedge \forall b^\beta.\mathrm{dom}\, fb \to q(o_0, b, o_1)(fb)\big)$

(22)
$$;sub \leadsto \lambda p^{\alpha \to (\beta \to \gamma) \to t}.\lambda q^{\alpha \times \beta \times (\beta \to \gamma) \to \delta \to t}.\lambda i^{\alpha}.\lambda o^{(\beta \to \gamma) \times (\beta \to \delta)}.pio_0$$
$$\wedge dom o_0 = dom o_1 \wedge \forall b^{\beta}.dom o_1 b \to q(i, b, o_0)(o_1 b)$$
$$: (S_{\alpha,(\beta \to \gamma) \times (\beta \to \delta)}/S_{\alpha \times \beta \times (\beta \to \gamma), \delta}) \backslash S_{\alpha, \beta \to \gamma}$$

We can now give a (summarized) derivation for (6), in (23).

(23) Derivation of (6). Let $\epsilon := e \times 1, \sigma := \epsilon \times v, \tau := \epsilon \to \sigma$ and $\omega := 1 \times \epsilon \times \tau$. Then:

every player
selects a pawn

$$\vdots$$

$$\begin{array}{c} S_{1,\tau} \quad \dfrac{;sub}{(S_{1,\tau \times (\epsilon \to v)}/S_{\omega,v}) \backslash S_{1,\tau}} \\ \hline \dfrac{S_{1,\tau \times (\epsilon \to v)}/S_{\omega,v}}{(S_{1,\tau \times (\epsilon \to v)})^{NP\omega}/(S_{\omega,v})^{NP\omega}} \, G \\ \hline \dfrac{((S_{1,\tau \times (\epsilon \to v)})^{NP\omega})^{NP\omega}/((S_{\omega,v})^{NP\omega})^{NP\omega}}{((S_{1,\tau \times (\epsilon \to v)})^{NP\omega})^{NP\omega}} \, G \quad ((S_{\omega,v})^{NP\omega})^{NP\omega} \end{array}$$

$$< \quad \begin{array}{c} \textit{he puts it} \\ \textit{on square one} \end{array}$$

$$\vdots$$

$$>$$

$$\vdots \; [close]$$

$$(S^{NP\omega})^{NP\omega}$$

Interpretation:

$$\lambda g^{\omega \to e}.\lambda h^{\omega \to e}.\exists F^{\tau \times (\epsilon \to v)}.dom\, F_0 = (\lambda o^{e \times 1}.player\, o_0)$$
$$\wedge (\forall u^{\epsilon}.dom\, F_0 u \to (pawn((F_0 u)_0)_0 \wedge select(u_0, ((F_0 u)_0)_0, (F_0 u)_1)))$$
$$\wedge dom\, F_0 = dom\, F_1 \wedge \forall a^{\epsilon}.dom\, F_1 a \to put(h(i, a, F_0), g(i, a, F_0), onsq1, (F_1 a))$$

We need to apply this formula to the resolutions for *it* and *he* respectively. The appropriate resolutions are shown in (24) and (25) respectively. They are both NRFs as defined in (19).

(24) $\lambda n^{1 \times (e \times 1) \times ((e \times 1) \to (e \times 1) \times v)}.(((n_1)_1 (n_1)_0)_0)_0$
(25) $\lambda n^{1 \times (e \times 1) \times ((e \times 1) \to (e \times 1) \times v)}.((n_1)_0)_0$

With those resolutions in place, the sentence is interpreted as shown below.[8]

$$\exists F^{\tau \times (\epsilon \to v)}.dom\, F_0 = (\lambda o^{e \times 1}.player\, o_0)$$
$$\wedge (\forall u^{\epsilon}.dom\, F_0 u \to (pawn((F_0 u)_0)_0 \wedge select(u_0, ((F_0 u)_0)_0, (F_0 u)_1)))$$
$$\wedge dom\, F_0 = dom\, F_1 \wedge \forall a^{\epsilon}.dom\, F_1 a \to put(a_0, ((F_0 a)_0)_0, onsq1, (F_1 a))$$

---

[8] As a result of the semantics assumed for *every*, the part of this formula corresponding to *every player selects a pawn* can be seen as a Skolemized version of (26).

(26) $\forall u^{e \times 1}.player\, u_0 \to \exists v^{(e \times 1) \times v}.pawn(v_0)_0 \wedge select(u_0, (v_0)_0, v_1)$

But since we have the Skolem function F_0, in the next conjunct the pronoun can be represented by $((F_0 a)_0)_0$, achieving the desired binding. A reviewer points out that this gives the analysis presented a certain resemblance to approaches that use epsilon terms to model indefinites and donkey pronouns (e.g. [18,24,28]), an observation for which I'm grateful.

This expresses the existence of a pair of functions, the first f of which maps every player to a pawn he chooses, and the second of which maps every player x to an event of x putting fx on square one.

The subordinating operator defined in (21), which is hypothesized to apply covertly in cases like (6), can also apparently be overt, as for example in (27).

(27) Every player chooses a pawn. He always puts it on square one.

And in fact, *always* in this sense seems to be just a special case of a variety of possible subordinators, as we can see from (28)–(29).

(28) Every player chooses a pawn. He usually puts it on square one.
(29) Every player chooses a pawn. He rarely puts it on square one.

The subordinating sentential conjunctions that apply in each of these cases are special instances of (30), where, as we have seen, det can at least be every (for *always*), most (for *usually*) or few (for *rarely*).

$$
\begin{aligned}
\text{(30)} \quad & \lambda p^{\alpha \to (\beta \to \gamma) \to t}.\lambda q^{\alpha \times \beta \times (\beta \to \gamma) \to \delta \to t}.\lambda i^{\alpha}.\lambda o^{(\beta \to \gamma) \times (\beta \to \delta)}.pio_0 \\
& \wedge \operatorname{dom} o_1 \subseteq \operatorname{dom} o_0 \wedge \mathsf{det}(\operatorname{dom} o_0)(\operatorname{dom} o_1) \\
& \wedge \forall b^{\beta}.\operatorname{dom} o_1 b \to q(i,b,o_0)(o_1 b)
\end{aligned}
$$

4.2   Constraints

Understood as a covert operator, the subordinator defined in (21) will cause vast overgeneration if allowed to apply too freely. It would, for example, allow the interpretation of *he* to covary with players in (31), but this is surely undesirable.

(31) ? Every player chooses a pawn. He has brown hair.

Empirical evidence is presented in [34] to show that a major constraint on this kind of quantificational subordination is the discourse relation that holds between the two sentences, where discourse relations are defined as in the framework of segmented discourse representation theory (SDRT, [22]). In that framework, the relation is taken to be *Narration* in the case of (6), but *Background* in the case of (31), for example. We can adopt this insight and take the discourse relation holding between sentences S_1 and S_2 as as a constraint on the applicability of a covert subordinator to the sentential conjunction coming between S_1 and S_2. I leave open the question of precisely the level at which this constraint should be stated.

5   Comparison with TTS

Existing discussion of quantificational subordination in TTS [31, 32] has focused almost exclusively on examples like (4). Now, the intended interpretations of (5) and (6) can certainly be *represented* in TTS, as shown in (32) and (33) respectively.[9]

---

[9] See [30, 33] for discussion of generalized quantifiers like Most in TTS.

(32)
$$(\Sigma f : (\Pi v : (\Sigma x : e)\text{WOMAN}(x)))$$
$$(\Sigma u : (\Sigma y : e)\text{BOOK}(y))\text{CHOOSE}(\pi_1(v), \pi_1(\pi_1(u)))$$
$$Most(\lambda x.\text{WOMAN}(x))(\lambda x.\text{WOMAN}(x) \times \text{READ}(x, \pi_1(\pi_1(f(x)))))$$

(33)
$$(\Sigma f : (\Pi v : (\Sigma x : e)\text{PLAYER}(x)))$$
$$(\Sigma u : (\Sigma y : e)\text{PAWN}(y))\text{SELECT}(\pi_1(v), \pi_1(\pi_1(u)))$$
$$(\Pi v : (\Sigma x : e)\text{PLAYER}(x))\text{PUT}(\pi_1(v), \pi_1(\pi_1(f(v))), \text{ONSQ1})$$

But the question is, how easily can those representations be derived compositionally? In the variety of TTS that has dealt most fully with the issues of compositionality and anaphora resolution, Dependent Type Semantics (DTS, [2,3,20,31,32]), pronouns are represented by @ terms, as defined in (34).

(34)
$$\frac{A : \textbf{type} \quad A \text{ true}}{(@ : A) : A} \; @$$

A term @ : A is well-typed in a context, then, iff it is provable that there is some term of type A in that context. Anaphoric resolution then amounts to replacing the @ term with some a : A at the point of type checking. In the version of DTS presented in [2,3], pronouns express @ terms that are functions from left contexts to entities, much like the system presented in this paper. An example is given in (35).[10]

(35)
$$it \rightsquigarrow \lambda P^{e \rightarrow \alpha \rightarrow \textbf{type}}.\lambda c^{\alpha}.P((@_i : \alpha \rightarrow e)(c))(c) : \text{S}/(\text{S}\backslash\text{NP})$$

So much for *it*, what about *they* or *of them*? A clue is given as to how this would work is given in ([31], p. 133), where it is stated that 'the type annotation of the @-term associated with *they* requires a predicate and a proof term of the cardinality condition'. However, no type annotation is actually given, so it is difficult to judge this claim. *Most* requires its first argument to be a predicate, i.e. (in DTS) something of type $e \rightarrow \textbf{type}$. It is reasonable to assume, then, that the @-term associated with *of them* should encode a function from left contexts to predicates. In the case of (5) the relevant @ term would therefore be as shown in (36).

(36) $@_j :$
$$\left(\begin{array}{c} (\Pi v : (\Sigma x : e)\text{WOMAN}(x))) \\ (\Sigma u : (\Sigma y : e)\text{BOOK}(y)) \\ \text{CHOOSE}(\pi_1(v), \pi_1(\pi_1(u))) \end{array} \right) \rightarrow e \rightarrow \textbf{type}$$

Without some equivalent of dom as discussed above, there is no way to get the right predicate, $\lambda x.\text{WOMAN}(x)$, out of the left context in (36).

Alternatively, one could eschew the functions-from-a-left-context approach to pronouns (as in [31,32]) and instead adopt a simpler perspective according to which *of them* would (presumably) be translated as $@_k : e \rightarrow \textbf{type}$. But then, in order to encode a subordinating conjunction to deal with cases like (6),

---

[10] Each @ term bears a unique index, which in the following examples are (arbitrarily) chosen as i, j, k.

something like dom is still needed. Standard sentential conjunction in this version of DTS is shown in (37); (38) shows an attempt to formulate a subordinating conjunction, but we don't know what type to put in in place of the question mark.

(37) $\lambda p.\lambda q.(\Sigma u : p)q$

(38) $\lambda p.\lambda q.(\Sigma u : p)(\Pi v :?)q$

Furthermore, the system presented in this paper benefits from a general definition of determiner meanings, as shown in Fig. 2 and discussed in Sect. 3.1. In contrast, while there has been some work on constructive generalized quantifiers appropriate for TTS [30,33], this is still at the ad-hoc, case-by-case stage, and there has been no discussion of monotone-decreasing determiners, for example.

6   Discussion and Future Work

At face value, many examples of anaphoric dependencies look like they depend on functional relationships established in discourse. We have shown that quite some progress in capturing those anaphoric dependencies can be made by taking that impression seriously, i.e. by having sentences denote functions and allowing those functions to serve as pronominal antecedents. We hope to have shown that this is a viable alternative to placeholders like sets of assignment functions, from which those functions have to be extracted.

One obvious next place to look for applications of this approach is in the treatment of 'paycheck' pronouns; for example, the interpretation of (39) according to which the second sentence is interpreted as equivalent to *every fourth grade boy hates his (own) mother.*

(39) Every third grade boy loves his mother. But every fourth grade boy hates her.   [16]

Once again, the interpretation of the second sentence gives the impression of depending on a functional relationship established in the first, namely between people and their mothers. And, in fact, many accounts of paycheck pronouns do in fact take that approach, either by saying that the relevant function is contextually salient [7] or, in a recent TTS analysis [32], that it is introduced by the presupposition of the possessive pronoun in the first sentence.

Another obvious avenue for extension is the phenomenon of modal subordination, as exemplified by (40).

(40) A wolf might come in. It would eat you first.   [26]

Of course, this would require an account of modality, which has not been offered yet.

Acknowledgements. This research is supported by an Early Career Fellowship from the Leverhulme Trust.

References

1. Barwise, J., Cooper, R.: Generalized quantifiers and natural language. Linguist. Philos. **4**(2), 159–219 (1981)
2. Bekki, D.: Representing anaphora with dependent types. In: Asher, N., Soloviev, S. (eds.) LACL 2014. LNCS, vol. 8535, pp. 14–29. Springer, Heidelberg (2014). https://doi.org/10.1007/978-3-662-43742-1_2
3. Bekki, D., Mineshima, K.: Context-passing and underspecification in dependent type semantics. In: Chatzikyriakidis, S., Luo, Z. (eds.) Modern Perspectives in Type-Theoretical Semantics. SLP, vol. 98, pp. 11–41. Springer, Cham (2017). https://doi.org/10.1007/978-3-319-50422-3_2
4. van den Berg, M.: Some aspects of the internal structure of discourse. Ph.D. thesis, University of Amsterdam (1996)
5. Brasoveanu, A.: Structured nominal and modal reference. Ph.D. thesis, Rutgers University (2007)
6. Chatzikyriakidis, S., Luo, Z. (eds.): Modern Perspectives in Type-Theoretical Semantics. SLP, vol. 98. Springer, Cham (2017). https://doi.org/10.1007/978-3-319-50422-3
7. Cooper, R.: The interpretation of pronouns. In: Heny, F., Schnelle, H. (eds.) Syntax and Semantics, vol. 10, pp. 61–92. Academic Press, New York (1979)
8. Cooper, R.: Records and record types in semantic theory. J. Logic. Comput. **15**(2), 99–112 (2005)
9. Elbourne, P.: Situations and Individuals. MIT Press, Cambridge (2005)
10. Geach, P.T.: Reference and generality. In: Contemporary Philosophy, Cornell University Press, Ithaca (1962)
11. Gotham, M.: A model-theoretic reconstruction of type-theoretic semantics for anaphora. In: Foret, A., Muskens, R., Pogodalla, S. (eds.) FG 2017. LNCS, vol. 10686, pp. 37–53. Springer, Heidelberg (2018). https://doi.org/10.1007/978-3-662-56343-4_3
12. Groenendijk, J., Stokhof, M.: Dynamic predicate logic. Linguist. Philos. **14**(1), 39–100 (1991)
13. Haug, D.T.T.: Partial dynamic semantics for anaphora. J. Seman. **31**, 457–511 (2014)
14. Heim, I.: The semantics of definite and indefinite nouns phrases. Ph.D. thesis, University of Massachussetts, Amherst (1982)
15. Jacobson, P.: Towards a variable-free semantics. Linguist, Philos. **22**(2), 117–184 (1999)
16. Jacobson, P.: Compositional Semantics. Oxford University Press, Oxford (2014)
17. Kamp, H.: A theory of truth and semantic representation. In: Groenendijk, J., Janssen, T., Stokhof, M. (eds.) Formal Methods in the Study of Language, pp. 277–322. Mathematisch Centrum, Amsterdam (1981)
18. Kempson, R., Meyer-Viol, W., Gabbay, D.: Dynamic Syntax. Blackwell, Oxford (2001)
19. Keshet, E.: Dynamic update anaphora logic. J. Seman. **35**, 263–303 (2018). https://doi.org/10.1093/jos/ffx020
20. Kinoshita, E., Mineshima, K., Bekki, D.: Coercion as proof search in dependent type semantics. Oslo Stud. Lang. **10**(2), 143–162 (2018)
21. Kratzer, A.: Situations in natural language semantics. In: Zalta, E.N. (ed.) The Stanford Encyclopedia of Philosophy. Metaphysics Research Lab, Stanford University, Summer 2019 (2019). https://plato.stanford.edu/archives/sum2019/entries/situations-semantics/

22. Lascarides, A., Asher, N.: Segmented discourse representation theory. In: Bunt, H., Muskens, R. (eds.) Computing Meaning, pp. 87–124. No. 83 in Studies in Linguistics and Philosophy, Springer, Dordrecht (2008). https://doi.org/10.1007/978-1-4020-5958-2_5

23. Martin-Löf, P.: An intuitionistic theory of types: predicative part. In: Rose, H., Shepherdson, J. (eds.) Logic Colloquium 1973, pp. 73–118. No. 80 in Studies in Logic and the Foundations of Mathematics, North-Holland, Amsterdam (1975)

24. Peregrin, J., von Heusinger, K.: Dynamic semantics with choice functions. In: Kamp, H., Partee, B. (eds.) Context-Dependence in the Analysis of Linguistic Meaning, pp. 255–274. Elsevier, Amsterdam (2004)

25. Ranta, A.: Type-Theoretical Grammar. No. 1 in Indices, Oxford University Press, Oxford (1994)

26. Roberts, C.: Modal subordination, anaphora and distributivity, Ph.D. thesis, University of Massachusetts at Amherst (1987)

27. Steedman, M.: The Syntactic Process. MIT Press, Cambridge (2000)

28. Steedman, M.: Taking Scope. MIT Press, Cambridge (2012)

29. Sundholm, G.: Proof theory and meaning. In: Gabbay, D., Guenther, F. (eds.) Handbook of philosophical logic. Synthese Library (Studies in Epistemology, Logic, Methodology, and Philosophy of Science), vol. 166, pp. 471–506. Springer, Dordrecht (1986). https://doi.org/10.1007/978-94-009-5203-4_8

30. Sundholm, G.: Constructive generalized quantifiers. Synthese **79**(1), 1–12 (1989)

31. Tanaka, R., Mineshima, K., Bekki, D.: On the interpretation of dependent plural anaphora in a dependently-typed setting. In: Kurahashi, S., Ohta, Y., Arai, S., Satoh, K., Bekki, D. (eds.) JSAI-isAI 2016. LNCS (LNAI), vol. 10247, pp. 123–137. Springer, Cham (2017). https://doi.org/10.1007/978-3-319-61572-1_9

32. Tanaka, R., Mineshima, K., Bekki, D.: Paychecks, presupposition, and dependent types. EasyChair Preprint no. 215 (2018). 10.29007/qw7n

33. Tanaka, R., Nakano, Y., Bekki, D.: Constructive generalized quantifiers revisited. In: Nakano, Y., Satoh, K., Bekki, D. (eds.) JSAI-isAI 2013. LNCS (LNAI), vol. 8417, pp. 115–124. Springer, Cham (2014). https://doi.org/10.1007/978-3-319-10061-6_8

34. Wang, L., McCready, E., Asher, N.: Information dependency in quantificational subordination. In: von Heusinger, K., Turner, K. (eds.) Where Semantics Meets Pragmatics, pp. 267–306. Elsevier (2003)

Undecidability of a Newly Proposed Calculus for CatLog3

Max Kanovich[1,4,5], Stepan Kuznetsov[2,4(✉)], and Andre Scedrov[3,4]

[1] University College London, London, UK
m.kanovich@ucl.ac.uk
[2] Steklov Mathematical Institute of the RAS, Moscow, Russia
sk@mi-ras.ru
[3] University of Pennsylvania, Philadelphia, USA
scedrov@math.upenn.edu
[4] National Research University Higher School of Economics, Moscow, Russia
{mkanovich,slkuznetsov,ascedrov}@hse.ru
[5] Institute of Oriental Studies of the RAS, Languages Department, Moscow, Russia

Abstract. In his recent papers "Parsing/theorem-proving for logical grammar CatLog3" and "A note on movement in logical grammar", Glyn Morrill proposes a new substructural calculus to be used as the basis for the categorial grammar parser CatLog3. In this paper we prove that the derivability problem for a fragment of this calculus is algorithmically undecidable.

Keywords: Categorial grammar · Contraction rule · Undecidability

1 Introduction

In his recent papers [31,32], Glyn Morrill proposes a new substructural calculus, to be used as the basis for the categorial grammar parser CatLog3. As the first step on the road of investigating algorithmic properties of the new Morrill's system, in this paper we shall prove that the derivability problem for a fragment of this calculus is algorithmically undecidable.

The source of undecidability is the contraction rule. In Morrill's systems, however, contraction appears in a very non-standard form. Moreover, the contraction rule presented in Morrill's new papers significantly differs from other ones, therefore, earlier undecidability proofs [12,14,23] do not work for this new version of contraction rule. Thus, a new technique should be invented, and we do that in the present paper.

The idea of categorial grammars goes back to Ajdukiewicz [2] and Bar-Hillel [3]. The version of categorial grammars used by Morrill is an extension of Lambek categorial grammars [22]. In a categorial grammar, each word (lexeme) of the language is given one or several syntactic categories (types), which are formulae of a specific logical system, an extension of the Lambek calculus.

© Springer-Verlag GmbH Germany, part of Springer Nature 2019
R. Bernardi et al. (Eds.): FG 2019, LNCS 11668, pp. 67–83, 2019.
https://doi.org/10.1007/978-3-662-59648-7_5

Parsing with categorial grammars, that is, checking whether a sentence is considered correct according to the grammar, reduces to checking derivability in the logical system involved. Namely, a sequence of words $a_1 \ldots a_n$ is accepted by the grammar if and only if there exist such formulae $A_1, \ldots, A_n$ that, for each i, A_i is one of the syntactic categories for a_i, and the *sequent* $A_1, \ldots, A_n \Rightarrow S$ is derivable. Here S is a designated syntactic category for grammatically correct sentences.

The Lambek calculus, which is used as the basis for categorial grammars, is a substructural logic and is closely related to Girard's linear logic [1,7]. In linear logic, formulae are treated as *resources,* thus, each formula should be used exactly once. This motivates the absence of the structural rules of contraction and weakening. Moreover, the Lambek calculus is also non-commutative, *i.e.,* does not include the rule of permutation (word order matters).

Sometimes, however, structural rules are allowed to be restored, in a restricted and controlled way, in order to treat subtle syntactic phenomena. One of such phenomena is *parasitic extraction,* which happens in phrases like "*the paper that John signed without reading.*" Here the dependent clause has two *gaps,* which we denote by []: "*John signed* [] *without reading* []," which should both be filled by the same "*the paper*" in order to obtain a complete sentence. In the logic, this is handled by the contraction rule in its non-local form:

$$\frac{\Gamma_1, !A, \Gamma_2, !A, \Gamma_3 \Rightarrow C}{\Gamma_1, !A, \Gamma_2, \Gamma_3 \Rightarrow C}$$

Here ! is the (sub)exponential modality, and the contraction rule is allowed to be applied to formulae of the form $!A$, and only to them.

Extension of the Lambek calculus with a subexponential modality which allows the non-local contraction rule formulated as presented above are undecidable [16]. In Morrill's systems, however, the contraction rule is presented in a rather non-standard form. The reason is in the usage of brackets which introduce controlled non-associativity. Brackets prevent the calculus from overgeneration, that is, from justifying grammatically incorrect sentences as correct ones. The contraction rule, as shown below, also essentially interacts with brackets. This makes standard undecidability proofs unapplicable to Morrill's systems, so new undecidability proofs are needed.

In order to make our examples more formal, we assign syntactic type N to noun phrases, like "*John*" or "*the paper,*" and S to grammatically correct sentences. Our dependent clause "*John signed without reading*" receives type $S \,/\, !N$, meaning a syntactic object which lacks a noun phrase in order to become a complete sentence ("*John signed the paper without reading the paper*"). The *subexponential modality* ! applied to N means that our noun phrase should be commutative (in order to find its place inside the sentence) and allow contraction (in order to fill both gaps).

Overgeneration is exhibited by the following example: * "*the paper that John signed and Pete ate a pie*" (the asterisk marks the sentence as ungrammatical). On one hand, this phrase is clearly ungrammatical, because of the irrelevant fragment *"Pete ate a pie"* in the dependent clause. On the other hand, being a sentence with one gap, *"John signed [] and Pete ate a pie"* receives the same type $S/!N$, which makes it equivalent to correct dependent clauses like *"John signed [] yesterday."* In order to address this issue, Morrill [26] and Moortgat [25] introduce brackets which embrace so-called islands, which, within our setting, cannot be penetrated by $!N$. In particular, and-coordination of sentences makes the result a bracketed island. Since phrases like * *"the paper that John left the office without reading"* are also ungrammatical, a *without*-clause also forms an island and should be embraced in brackets. This leads to Morrill's idea of handling parasitic extraction [27, Sect. 5.5]: in the dependent clause, there is one *principal*, or *host* gap, which should not be inside an island, and *parasitic* gaps, which reside in islands. Moreover, a parasitic gap can also be a host for its own "second-order" parasitic gaps.

Thus, the contraction rule should take one $!A$ from a bracket-embraced island and remove it, in the presence of another $!A$ outside the island. However, after that the bracketing should be somehow changed, in order to avoid another usage of the same island for parasitic gapping. This general idea, however, has different realisations in a number of works of Morrill and his co-authors [27–29,31,32]. In the next section we show the most recent approach [31,32], which essentially resembles the original construction from Morrill's 2011 book [27].

2   The Calculus

Morrill's calculus for CatLog3 [32] is quite involved, including up to 45 connectives. The metasyntax of sequents in this calculus is also rather non-standard, involving brackets and meta-operations for discontinuity. In this paper we consider its simpler fragment, involving only the multiplicative Lambek operations: left and right divisions ($\backslash$, $/$), multiplication ($\bullet$), and the unit ($\mathbf{I}$), brackets and bracket modalities ($\langle\rangle$, $[]^{-1}$), and the subexponential modality, and already for this fragment we show undecidability.

Notice that Morrill's system also includes Kleene star, axiomatised by means of an ω-rule. In Morrill's system, it is called "existential exponential" and denoted by "?". In the presence of the Kleene star the Lambek calculus is known to be at least Π_1^0-hard [5,20] and thus undecidable. Moreover, in the view of Kozen's [18] results on complexity of Horn theories of Kleene algebras, the complexity of the system including *both* Kleene star and the subexponential could potentially rise even higher, up to Π_1^1-completeness. Morrill, however, emphasizes the fact that in formulae used in categorial grammars designed for real languages the Kleene star never occurs with positive polarity. Thus, the ω-rule is never used, and the Kleene star does not incur problems with decidability. Thus, the only possible source of undecidability is the specific contraction rule for the subexponential. We consider a fragment of Morrill's system with this rule, which is sufficient to show undecidability.

Let us define the syntax of our fragment. Formulae will be built from variables (primitive types) $p, q, \ldots$ and the multiplicative unit constant $\mathbf{I}$ using three binary operations: $\backslash$ (left division), $/$ (right division), $\bullet$ (product), and three unary operation: $\langle\rangle$ and $[]^{-1}$ (bracket modalities) and $!$ (subexponential). Sequents (in Morrill's terminology, *h-sequents*) of $!^{\mathbf{s}}_{\mathbf{b1/2}}\mathbf{L}^*\mathbf{b}$ are expressions of the form $\Xi \Rightarrow A$, where A is a formula and Ξ is a complex metasyntactic structure which we call *meta-formula* (Morrill calls them *zones*). Meta-formulae are built from formulae using comma and brackets; also, formulae which are intended to be marked by the subexponential $!$, which allows permutation, are placed into special commutative areas called *stoups* (cf. [8,9,17]). Following Morrill [32], we define the notion of meta-formula along with two auxiliary notions, stoup and *tree term*, simultaneously.

- A stoup is a multiset of formulae: $\zeta = \{A_1, \ldots, A_n\}$. A stoup could be empty, the empty stoup is denoted by $\varnothing$.
- A tree term is either a formula or a bracketed expression of the form $[\Xi]$, where Ξ is a meta-formula.
- A meta-formula is an expression of the form $\zeta; \Gamma$, where ζ is a stoup and Γ is a linearly ordered sequence of tree terms. Here Γ could also be empty; the empty sequence is denoted by Λ.

We use comma both for concatenation of tree term sequences and for multiset union of stoups (Morrill uses $\uplus$ for the latter). Moreover, for adding one formula into a stoup we write ζ, A instead of $\zeta, \{A\}$.

Axioms and rules of $!^{\mathbf{s}}_{\mathbf{b1/2}}\mathbf{L}^*\mathbf{b}$ are as follows.

$$\overline{\varnothing; A \Rightarrow A} \ id$$

$$\frac{\zeta_1; \Gamma \Rightarrow B \quad \Xi(\zeta_2; \Delta_1, C, \Delta_2) \Rightarrow D}{\Xi(\zeta_1, \zeta_2; \Delta_1, C / B, \Gamma, \Delta_2) \Rightarrow D} \ /L \qquad \frac{\zeta; \Gamma, B \Rightarrow C}{\zeta; \Gamma \Rightarrow C / B} \ /R$$

$$\frac{\zeta_1; \Gamma \Rightarrow A \quad \Xi(\zeta_2; \Delta_1, C, \Delta_2) \Rightarrow D}{\Xi(\zeta_1, \zeta_2; \Delta_1, \Gamma, A \backslash C, \Delta_2) \Rightarrow D} \ \backslash L \qquad \frac{\zeta; A, \Gamma \Rightarrow C}{\zeta; \Gamma \Rightarrow A \backslash C} \ \backslash R$$

$$\frac{\Xi(\zeta; \Delta_1, A, B, \Delta_2) \Rightarrow D}{\Xi(\zeta; \Delta_1, A \bullet B, \Delta_2) \Rightarrow D} \ \bullet L \qquad \frac{\zeta_1; \Delta \Rightarrow A \quad \zeta_2; \Gamma \Rightarrow B}{\zeta_1, \zeta_2; \Delta, \Gamma \Rightarrow A \bullet B} \ \bullet R$$

$$\frac{\Xi(\zeta; \Delta_1, \Delta_2) \Rightarrow A}{\Xi(\zeta; \Delta_1, \mathbf{I}, \Delta_2) \Rightarrow A} \ \mathbf{IL} \qquad \overline{\varnothing; \Lambda \Rightarrow \mathbf{I}} \ \mathbf{IR}$$

$$\frac{\Xi(\zeta; \Delta_1, A, \Delta_2) \Rightarrow B}{\Xi(\zeta; \Delta_1, [\varnothing; []^{-1}A], \Delta_2) \Rightarrow B} \ []^{-1}L \qquad \frac{\varnothing; [\Xi] \Rightarrow A}{\Xi \Rightarrow []^{-1}A} \ []^{-1}R$$

$$\frac{\Xi(\zeta; \Delta_1, [\varnothing; A], \Delta_2) \Rightarrow B}{\Xi(\zeta; \Delta_1, \langle\rangle A, \Delta_2) \Rightarrow B} \ \langle\rangle L \qquad \frac{\Xi \Rightarrow A}{\varnothing; [\Xi] \Rightarrow \langle\rangle A} \ \langle\rangle R$$

$$\frac{\Xi(\zeta, A; \Gamma_1, \Gamma_2) \Rightarrow B}{\Xi(\zeta; \Gamma_1, !A, \Gamma_2) \Rightarrow B} \ !L \qquad \frac{\varnothing; !A \Rightarrow B}{\varnothing; !A \Rightarrow !B} \ !R$$

$$\frac{\Xi(\zeta; \Gamma_1, A, \Gamma_2) \Rightarrow B}{\Xi(\zeta, A; \Gamma_1, \Gamma_2) \Rightarrow B} \; !P \qquad \frac{\Xi(\zeta, A; \Gamma_1, [A; \Gamma_2], \Gamma_3) \Rightarrow B}{\Xi(\zeta, A; \Gamma_1, [\varnothing; [\varnothing; \Gamma_2]], \Gamma_3) \Rightarrow B} \; !C$$

Morrill [31,32] does not give any particular name to his calculus. In this paper, we denote our fragment by $!^s_{b1/2}\mathbf{L}^*\mathbf{b}$. Here "b" stands for "bracketed," and the decorations of ! mean the following. The superscript "s" means that the right rule for ! is in the style of soft and light linear logic [10,15,21], allowing, in particular, only one $!A$ in the left-hand side. The subscript "b1/2" means that contraction operates brackets, using single bracketing in the premise and double bracketing in the conclusion.

In his older paper [29], Morrill uses another form of contraction rule, which in our notation looks like

$$\frac{\Xi(\zeta, A; \Gamma_1, [A; \Gamma_2], \Gamma_3) \Rightarrow B}{\Xi(\zeta, A; \Gamma_1, \Gamma_2, \Gamma_3) \Rightarrow B}$$

Thus, this system could be called $!^s_{b1/0}\mathbf{L}^*\mathbf{b}$, in our notations. For the system with this sort of contraction, undecidability was established in [14]. The new contraction rule of Morrill [31,32], however, significantly differs from the old contraction rule, and the undecidability proof from [14] does not work for Morrill's new system. Thus, undecidability becomes a separate issue and we address it in this paper.

For convenience, we use the following derivable *dereliction* rule

$$\frac{\Xi(\zeta; \Gamma_1, A, \Gamma_2) \Rightarrow B}{\Xi(\zeta; \Gamma_1, !A, \Gamma_2) \Rightarrow B} \; !D$$

which is actually consecutive application of $!P$ and $!L$:

$$\frac{\dfrac{\Xi(\zeta; \Gamma_1, A, \Gamma_2) \Rightarrow B}{\Xi(\zeta, A; \Gamma_1, \Gamma_2) \Rightarrow B} \; !P}{\Xi(\zeta; \Gamma_1, !A, \Gamma_2) \Rightarrow B} \; !L$$

Notice that in Morrill's calculus [31,32] there is no cut rule. Thus, the question of cut-elimination is transformed into the question of admissibility of cut, proving which is marked in [32] as an ongoing work by O. Valentín. Since the calculus considered in [31,32] does not include cut, our fragment, which uses only a restricted set of connectives and consists of the corresponding inference rules, is a conservative fragment of the complete system [32]. Namely, for sequents in the restricted language, derivability in the fragment is equivalent to derivability in the big system. Therefore, undecidability for $!^s_{b1/2}\mathbf{L}^*\mathbf{b}$ (Theorem 3 below) yields undecidability for the whole system also.

Using $!^s_{b1/2}\mathbf{L}^*\mathbf{b}$, one can analyze our example *"the paper that John signed without reading"* in the following way, simplyfing Morrill's analysis [32]. Assign the following syntactic types to words:

$$the \triangleright N \,/\, CN \qquad\qquad likes,\ signed \triangleright (\langle\rangle N \setminus S) \,/\, N$$

$$man,\ paper \triangleright CN \qquad without \triangleright ([]^{-1}(((\langle\rangle N \setminus S) \setminus (\langle\rangle N \setminus S))) \,/\, (\langle\rangle N \setminus S)$$

$$reading \triangleright (\langle\rangle N \setminus S) \,/\, N \qquad who,\ that \triangleright ([]^{-1}[]^{-1}(CN \setminus CN)) \,/\, (S \,/\, !N)$$

$$John \triangleright \langle\rangle N$$

Here N stands for "noun phrase," CN states for "common noun" (without an article), and S stands for "sentence".

In order to parse this sentence in this grammar, one first needs to impose the bracketing structure on it. This is done in the following way:

the paper [[*that* [*John*] *signed* [[*without reading*]]]].

Indeed, in Morrill's CatLog categorial grammar the subject group and the *without*-clause form islands, and the *that*-clause forms a strong island, embraced by double brackets. Moreover, we also have to double-bracket our *without*-clause (make it a "strong island"), since it will be used for parasitic extraction. Each pair of brackets has its own stoup, which is originally empty. Unfortunately, in CatLog the bracketed structure is required as an input from the user (while it is of course not part of the original sentence). Morrill et al. [30], however, provide an algorithm for automated induction (guessing) of the bracketed structure, for a small fragment of the CatLog grammar (in particular, without subexponential).

With the bracketing shown above, the corresponding sequent is derived in $!^{s}_{b1/2}\mathbf{L^*b}$ as shown in Fig. 1.

At the request of one of the referees, we discuss the following example, which is used by Morrill [31] to motivate the changes made in the contraction rule from $\mathbf{b1/0}$ to $\mathbf{b1/2}$ (see above). This example features an incorrect noun phrase, * *"the man who likes,"* analysed with two gaps in the dependent clause: * *"the man who* [] *likes* []*."* (Asterisks denote ungrammaticality.) The intended semantics (and the correct version of the phrase) here is *"the man who likes himself."* In $!^{s}_{b1/0}\mathbf{L^*b}$, however, * *"the man who* [] *likes* []*,"* with brackets imposed as *"the man* [[*who likes*]]*,"* is parsed as follows. First one derives the sequent $\varnothing; (\langle\rangle N \setminus S) \,/\, N \Rightarrow S \,/\, !N$, which (ungrammatically) treats *"likes"* as a dependent clause with two gaps, a host one for the object and a parasitic one for the subject:

$$\dfrac{\dfrac{\dfrac{\dfrac{\dfrac{\dfrac{\dfrac{\dfrac{\dfrac{\dfrac{\varnothing; N \Rightarrow N}{[\varnothing; N] \Rightarrow \langle\rangle N} \quad \varnothing; S \Rightarrow S}{\varnothing; [\varnothing; N], \langle\rangle N \setminus S \Rightarrow S}}{\varnothing; [\varnothing; N], (\langle\rangle N \setminus S) \,/\, N, N \Rightarrow S}}{\varnothing; [N; \Lambda], (\langle\rangle N \setminus S) \,/\, N, N \Rightarrow S}}{N; [N; \Lambda], (\langle\rangle N \setminus S) \,/\, N, N \Rightarrow S}}{N; (\langle\rangle N \setminus S) \,/\, N \Rightarrow S}}{\varnothing; (\langle\rangle N \setminus S) \,/\, N, !N \Rightarrow S}}{\varnothing; (\langle\rangle N \setminus S) \,/\, N \Rightarrow S \,/\, !N}$$

Fig. 1. Derivation for *"the paper that John signed without reading"* (cf. [32, Fig. 24])

Here the whole subject island is introduced by $!C$ (in its **b1/0** version, with $\Gamma_2 = \Lambda$) as a parasitic extraction site. Next, one finishes the derivation as it is done in Fig. 1 and obtains

$$\varnothing; N \,/\, CN, CN, [\varnothing; [\varnothing; ([]^{-1}[]^{-1}(CN \setminus CN) \,/\, (S \,/\, !N), (\langle\rangle N \setminus S) \,/\, N]] \Rightarrow N.$$

With the new, **b1/2** contraction rule, this derivation of *"the man [[who likes]]"* becomes impossible. However, there still exists a way to derive *"the man who likes,"* if the user imposes the following weird bracketing: *"the man [[who [[ ]] likes]]."* This bracketing explicitly creates an empty strong, double-bracketed island as the subject of the dependent clause, and the $!C$ rule transforms it into a single-bracketed one. (In other parts, the derivation is similar to the one presented above.) In one of the reviews, the referee asks whether one can consider a system where empty brackets are explicitly disallowed, and whether our undecidability proof is still valid for this system. This constraint, however, is tightly connected with the Lambek's antecedent non-emptiness restriction. It appears that reconciling this constraint with (sub)exponential modalities raises certain issues with keeping good proof-theoretic properties of the system, such as cut elimination and substitution [11, 13]. We accurately formulate these questions in the "Future Work" section and leave them as open problems for future research.

3   The Bracket-Free System and the π Projection

In this section we define $!L^*$, a system without brackets and with a full-power exponential modality. This is a more well-known system, and it is simpler from the logical point of view. We shall need $!L^*$ inside our undecidability proof in Sect. 4. In this section we define a projection that maps derivability in $!^s_{b1/2}L^*b$ to derivability in $!L^*$. This projection is similar to the bracket-forgetting projection in [14].

Formulae of $!L^*$ are defined similary to the ones of $!^s_{b1/2}L^*b$, but without bracket modalities ($\langle\rangle$ and $[]^{-1}$). Sequents of $!L^*$ have a simpler structure, and are expressions of the form $\Gamma \Rightarrow A$, where A is a formula and Γ is a linearly ordered sequence of formulae. Axioms and inference rules of $!L^*$ are as follows.

$$\overline{A \Rightarrow A} \;\; id$$

$$\frac{\Gamma \Rightarrow B \quad \Delta_1, C, \Delta_2 \Rightarrow D}{\Delta_1, C \,/\, B, \Gamma, \Delta_2 \Rightarrow D} \;/L \qquad \frac{\Gamma, B \Rightarrow C}{\Gamma \Rightarrow C \setminus B} \;/R$$

$$\frac{\Gamma \Rightarrow A \quad \Delta_1, C, \Delta_2 \Rightarrow D}{\Delta_1, \Gamma, A \setminus B, \Delta_2 \Rightarrow D} \;\backslash L \qquad \frac{A, \Gamma \Rightarrow C}{\Gamma \Rightarrow A \setminus C} \;/R$$

$$\frac{\Delta_1, A, B, \Delta_2 \Rightarrow D}{\Delta_1, A \bullet B, \Delta_2 \Rightarrow D} \;\bullet L \qquad \frac{\Delta \Rightarrow A \quad \Gamma \Rightarrow B}{\Delta, \Gamma \Rightarrow A \bullet B} \;\bullet R$$

$$\frac{\Delta_1, \Delta_2 \Rightarrow A}{\Delta_1, \mathbf{I}, \Delta_2 \Rightarrow A} \ IL \qquad \frac{}{\Lambda \Rightarrow \mathbf{I}} \ IR$$

$$\frac{\Gamma_1, A, \Gamma_2 \Rightarrow B}{\Gamma_1, !A, \Gamma_2 \Rightarrow B} \ !L \qquad \frac{!A_1, \ldots, !A_n \Rightarrow B}{!A_1, \ldots, !A_n \Rightarrow !B} \ !R \qquad \frac{\Gamma_1, \Gamma_2 \Rightarrow B}{\Gamma_1, !A, \Gamma_2 \Rightarrow B} \ !W$$

$$\frac{\Gamma_1, !A, \Gamma_2, !A, \Gamma_3 \Rightarrow B}{\Gamma_1, !A, \Gamma_2, \Gamma_3 \Rightarrow B} \ !C_1 \qquad \frac{\Gamma_1, !A, \Gamma_2, !A, \Gamma_3 \Rightarrow B}{\Gamma_1, \Gamma_2, !A, \Gamma_3 \Rightarrow B} \ !C_2$$

$$\frac{\Gamma \Rightarrow A \quad \Delta_1, A, \Delta_2 \Rightarrow D}{\Delta_1, \Gamma, \Delta_2 \Rightarrow D} \ cut$$

Notice that, unlike $!^{\mathbf{s}}_{\mathbf{b}1/2}\mathbf{L}^*\mathbf{b}$, here cut is included as an official rule of the system. However, here the cut rule is eliminable by a standard technique by using the mix rule.

Proposition 1. *Any sequent derivable in* $!\mathbf{L}^*$ *is derivable without using the cut rule.*

This proof of cut elimination is explained, for example, in [16], where $!\mathbf{L}^*$ acts as a specific case of $\mathbf{SMALC}_\Sigma$, an extension of the multiplicative-additive Lambek calculus with a family of subexponential modalities.

In our version of $!\mathbf{L}^*$, contraction rules ($!C_1$ and $!C_2$) are non-local (cf. [16]), and permutation rules of the following form

$$\frac{\Gamma_1, \Gamma_2, !A, \Gamma_3 \Rightarrow B}{\Gamma_1, !A, \Gamma_2, \Gamma_3 \Rightarrow B} \ !P_1 \qquad \frac{\Gamma_1, !A, \Gamma_2, \Gamma_3 \Rightarrow B}{\Gamma_1, \Gamma_2, !A, \Gamma_3 \Rightarrow B} \ !P_2$$

are derivable using non-local contraction and weakening:

$$\frac{\dfrac{\Gamma_1, \Gamma_2, !A, \Gamma_3 \Rightarrow B}{\Gamma_1, !A, \Gamma_2, !A, \Gamma_3 \Rightarrow B} \ !W}{\Gamma_1, !A, \Gamma_2, \Gamma_3 \Rightarrow B} \ !C_1 \qquad \frac{\dfrac{\Gamma_1, !A, \Gamma_2, \Gamma_3 \Rightarrow B}{\Gamma_1, !A, \Gamma_2, !A, \Gamma_3 \Rightarrow B} \ !W}{\Gamma_1, \Gamma_2, !A, \Gamma_3 \Rightarrow B} \ !C_2$$

Next, we define a translation from $!^{\mathbf{s}}_{\mathbf{b}1/2}\mathbf{L}^*\mathbf{b}$ to $!\mathbf{L}^*$, which is actually a forgetting projection, erasing all brackets and bracket modalities, and also translating stoups into plain sequences of !-formulae. We denote this projection by π and define it in the following recursive way.

– For a formula A, its projection $\pi(A)$ is defined as follows:

$$\pi(p) = p \text{ for any variable } p; \qquad \pi(\mathbf{I}) = \mathbf{I};$$
$$\pi(A \backslash B) = \pi(A) \backslash \pi(B); \qquad \pi(B / A) = \pi(B) / \pi(A);$$
$$\pi(A \bullet B) = \pi(A) \bullet \pi(B); \qquad \pi(!A) = !\pi(A);$$
$$\pi(\langle \rangle A) = \pi([]^{-1}A) = \pi(A).$$

- For a stoup $\zeta = \{A_1, \ldots, A_n\}$, its π-projection is the sequence of formulae $!\pi(A_1), \ldots, !\pi(A_n)$. Since in $!\mathbf{L}^*$ we have permutation rules for !-formulae, the order does not matter. The π-projection of an empty stoup is the empty sequence Λ.
- For a tree term there are two cases. If it is a formula, A, then its π-projection is $\pi(A)$. If the tree term is of the form $[\Xi]$, where Ξ is a meta-formula, then its π-projection is $\pi(\Xi)$ (as defined below).
- For a meta-formula of the form $\zeta; \Upsilon_1, \ldots, \Upsilon_k$, where Υ_i are tree terms, its π-projection is $\pi(\zeta), \pi(\Upsilon_1), \ldots, \pi(\Upsilon_k)$.

Proposition 2. *If $\Xi \Rightarrow A$ is derivable in $!^s_{b1/2}\mathbf{L}^*\mathbf{b}$, then $\pi(\Xi) \Rightarrow \pi(A)$ is derivable in $!\mathbf{L}^*$.*

Proof. Proceed by induction on derivation; recall that it is cut-free by definition.

Axioms id and $\mathbf{IR}$ and rules $/R$, $\bullet L$, and $\mathbf{IL}$ of $!^s_{b1/2}\mathbf{L}^*\mathbf{b}$, translate exactly to the corresponding rules of $!\mathbf{L}^*$. The rules for bracket modalities ($\langle\rangle L$, $\langle\rangle R$, $[]^{-1}L$, $[]^{-1}R$) become trivial: after applying the π-projection, the conclusion of such a rule coincides with its premise. For the rules $\backslash R$, $\bullet R$, $/L$, and $\backslash L$ are translated to the corresponding rules in $!\mathbf{L}^*$, together with necessary permutations ($!P_{1,2}$) for !-formulae coming from the stoups. Finally, the !-rules of $!^s_{b1/2}\mathbf{L}^*\mathbf{b}$ translate to the corresponding rules of $!\mathbf{L}^*$: $!L$ becomes $!P_2$, $!R$ maps to $!R$, $!P$ becomes $!L$ together with $!P_1$, and $!C$ maps to $!C_1$.

Notice that the reverse implication does not hold, which can be shown by analysis of our examples for brackets, like *"the paper that John signed and Pete ate a pie."*

4   Undecidability Proof

In this section we prove undecidability of the derivability problem in $!^s_{b1/2}\mathbf{L}^*\mathbf{b}$.

Theorem 3. *The derivability problem in $!^s_{b1/2}\mathbf{L}^*\mathbf{b}$ is undecidable, more precisely, Σ^0_1-complete.*

The general outline of our proof is rather standard, following the ideas of Lincoln et al. [23]: encoding of semi-Thue systems in $!^s_{b1/2}\mathbf{L}^*\mathbf{b}$. Maintaining the correct bracket structure, however, makes the encoding more involved and requires some technical tricks.

A *semi-Thue system* [34] over an alphabet Σ is a set of pairs of words over Σ, called *rewriting rules* and written as $x_1 \ldots x_m \to y_1 \ldots y_k$ ($k, m \geq 0$, $x_i, y_i \in \Sigma$). A rewriting sequence in a semi-Thue system S is a sequence of words $w_1, w_2, \ldots, w_N$, in which each word w_ℓ, starting from the second one, is obtained from the previous word $w_{\ell-1}$ by applying a rewriting rule as follows:

$$w_{\ell-1} = u x_1 \ldots x_m v \to u y_1 \ldots y_k v = w_\ell,$$

where $x_1 \ldots x_m \to y_1 \ldots y_k$ is a rewriting rule of $\mathcal{S}$ and u, v are arbitrary words. If there exists a rewriting sequence $w_1 \to w_2 \to \ldots \to w_N$ in $\mathcal{S}$, we say that w_N is derivable from w_1 in $\mathcal{S}$.

A famous result by Markov [24] and Post [33] shows that the derivability problem for semi-Thue systems is undecidable; more precisely, it is Σ_1^0-complete (that is, the membership problem for any recursively enumerable language can be reduced to the derivability problem in semi-Thue systems). Moreover, the problem of derivability of a word w from a one-letter word s, like in Chomsky's [6] type-0 grammars, is also undecidable. This can be shown by the following reduction: for arbitrary words w_1 and w_N, checking derivability of w_N from w in a semi-Thue system $\mathcal{S}$ is equivalent to checking derivability of w_N from the one-letter word s in the semi-Thue system $\mathcal{S}$ extended by a new symbol s and a new rewriting rule $s \to w_1$.

Let us proceed with our encoding of semi-Thue systems in $!^s_{\mathbf{b}1/2}\mathbf{L}^*\mathbf{b}$. Let

$$\mathcal{A}_\mathcal{S} = \{(x_1 \bullet \ldots \bullet x_m)/(y_1 \bullet \ldots \bullet y_k) \mid$$
$$x_1 \ldots x_m \to y_1 \ldots y_k \text{ is a rewriting rule of } \mathcal{S}\}.$$

If the word $x_1 \ldots x_m$ is empty, then $x_1 \bullet \ldots \bullet x_m$ is replaced by $\mathbf{I}$; the same for $y_1 \ldots y_k$.

Let $\mathcal{A}_\mathcal{S} = \{A_1, \ldots, A_n\}$ (the order does not matter). For each A_i let

$$Z_i = []^{-1}(!A_i \bullet \langle\rangle\langle\rangle\mathbf{I})$$

and define the following two sets of formulae (further they will be considered as multisets and used in the stoup):

$$\mathcal{Z}_\mathcal{S} = \{!Z_1, \ldots, !Z_n\};$$
$$\mathcal{X}_\mathcal{S} = \{\mathbf{I}/!Z_1, \ldots, \mathbf{I}/!Z_n, \mathbf{I}/(\langle\rangle\langle\rangle\mathbf{I})\}.$$

Finally, consider the following linearly ordered sequence of formulae:

$$\Gamma_\mathcal{S} = !A_1, \ldots, !A_n.$$

The intuition behind Z_i is as follows and is best understood when reading simultaneously with the formal proof of the $1 \Rightarrow 2$ implication in Theorem 4 below. In the sequent, we keep a special empty tree-term with double bracketing, $[\varnothing; [\varnothing; \Lambda]]$, which is used as the "landing zone" for Z_i. Double brackets, with empty stoups, allow the usage of Morrill's contraction rule, $!C$. Applying this rule (we trace the derivation tree from bottom to top) destroys one pair of brackets and puts $!Z_i$, taken from the stoup, inside. Dereliction removes the $!$, and the bracket modality inside Z_i destroys the second pair of brackets around it. Now we have $!A_i$ and $\langle\rangle\langle\rangle\mathbf{I}$. The former, by $!L$ and $!P$, is put to an arbitrary place of the antecedent, allowing application of a rewriting rule of the semi-Thue system $\mathcal{S}$. The latter restores the landing zone which was destroyed by contraction, and leaves a configuration which is ready for the next reduction step. Finally, formulae from $\mathcal{X}_\mathcal{S}$ are used for garbage collection on the top of the derivation.

This gives a translation of semi-Thue derivations to $!^s_{b1/2}L^*b$ ones. The backwards translation, from $!^s_{b1/2}L^*b$ derivations back to $\mathcal{S}$, is performed via the π-projection. This projection trivialises everything connected to brackets, and the resulting sequent, derivable in $!L^*$ by Proposition 2, happens to be $!L^*$-equivalent to the standard encoding as in [23]. Thus, the fact that its derivability yields the corresponding derivability in $\mathcal{S}$ is proved by the good old argument. Notice that in our reasoning we never use the cut rule: semi-Thue derivations are encoded by cut-free derivations in $!^s_{b1/2}L^*b$, the π-projection maps them onto cut-free derivations in $!L^*$, and they are mapped back onto semi-Thue derivations.

The idea described above is formalised by the following theorem, which serves as the principal technical lemma for Theorem 3.

Theorem 4. *The following three statements are equivalent:*

1. *the word $a_1 \ldots a_n$ is derivable from s in the semi-Thue system $\mathcal{S}$;*
2. *the sequent $\mathcal{X}_\mathcal{S}, \mathcal{Z}_\mathcal{S}; [\varnothing; [\varnothing; \Lambda]], a_1, \ldots, a_n \Rightarrow s$ is derivable in $!^s_{b1/2}L^*b$;*
3. *the sequent $\Gamma_\mathcal{S}, a_1, \ldots, a_n \Rightarrow s$ is derivable in $!L^*$.*

Proof. We establish the equivalence by proving round-robin implications: $1 \Rightarrow 2 \Rightarrow 3 \Rightarrow 1$.

$\boxed{1 \Rightarrow 2}$ This part of the proof formalises the idea we explained just before formulating Theorem 4. Proceed by induction on the length of the rewriting sequence. Induction base is $n = 1$, $a_1 = s$, and the necessary sequent, $\mathcal{X}_\mathcal{S}, \mathcal{Z}_\mathcal{S}; [\varnothing; [\varnothing; \Lambda]], s \Rightarrow s$, is derived as follows:

$$
\cfrac{
\cfrac{
\cfrac{\varnothing; \Lambda \Rightarrow \mathbf{I}}{\varnothing; [\varnothing; \Lambda] \Rightarrow \langle\rangle\mathbf{I}} \langle\rangle R
}{\varnothing; [\varnothing; [\varnothing; \Lambda]] \Rightarrow \langle\rangle\langle\rangle\mathbf{I}} \langle\rangle R
\qquad
\cfrac{
\cfrac{
\cfrac{
\cfrac{\varnothing;!Z_1 \Rightarrow !Z_1 \quad \ldots \quad \varnothing;!Z_n \Rightarrow !Z_n \quad \cfrac{\varnothing; s \Rightarrow s}{\varnothing;\mathbf{I},\ldots,\mathbf{I}, s \Rightarrow s} \text{IL } (n \text{ times})}{\varnothing; \mathbf{I}/!Z_1,!Z_1,\ldots,\mathbf{I}/!Z_n,!Z_n, s \Rightarrow s} /L \ (n \text{ times})
}{\mathbf{I}/!Z_1,\ldots,\mathbf{I}/!Z_n,!Z_1,\ldots,!Z_n; s \Rightarrow s} !P \ (2n \text{ times})
}{\mathbf{I}/!Z_1,\ldots,\mathbf{I}/!Z_n,!Z_1,\ldots,!Z_n;\mathbf{I}, s \Rightarrow s} \text{IL}
}{\mathbf{I}/!Z_1,\ldots,\mathbf{I}/!Z_n,!Z_1,\ldots,!Z_n;\mathbf{I}/(\langle\rangle\langle\rangle\mathbf{I}),[\varnothing;[\varnothing;\Lambda]], s \Rightarrow s} /L
}{\mathbf{I}/!Z_1,\ldots,\mathbf{I}/!Z_n,\mathbf{I}/(\langle\rangle\langle\rangle\mathbf{I}),!Z_1,\ldots,!Z_n;[\varnothing;[\varnothing;\Lambda]], s \Rightarrow s} !P
$$

For the induction step, we first establish derivability of the following "landing" rule:

$$
\cfrac{\mathcal{X}_\mathcal{S}, \mathcal{Z}_\mathcal{S}; [\varnothing; [\varnothing; \Lambda]], a_1, \ldots, a_i, A_j, a_{i+1}, \ldots, a_n \Rightarrow s}{\mathcal{X}_\mathcal{S}, \mathcal{Z}_\mathcal{S}; [\varnothing; [\varnothing; \Lambda]], a_1, \ldots, a_i, a_{i+1}, \ldots, a_n \Rightarrow s} \ land
$$

for any $A_j \in \mathcal{A}_\mathcal{S}$. This rule is derived as follows:

$$\frac{\mathcal{X}_\mathcal{S}, \mathcal{Z}_\mathcal{S}; [\varnothing; [\varnothing; \Lambda]], a_1, \ldots a_i, A_j, a_{i+1}, \ldots, a_n \Rightarrow s}{\mathcal{X}_\mathcal{S}, \mathcal{Z}_\mathcal{S}, A_j; [\varnothing; [\varnothing; \Lambda]], a_1, \ldots a_i, a_{i+1}, \ldots, a_n \Rightarrow s} \;!P$$

$$\frac{}{\mathcal{X}_\mathcal{S}, \mathcal{Z}_\mathcal{S}; !A_j, [\varnothing; [\varnothing; \Lambda]], a_1, \ldots a_i, a_{i+1}, \ldots, a_n \Rightarrow s} \;!L$$

$$\frac{}{\mathcal{X}_\mathcal{S}, \mathcal{Z}_\mathcal{S}; !A_j, [\varnothing; [\varnothing; \mathbf{I}]], a_1, \ldots a_i, a_{i+1}, \ldots, a_n \Rightarrow s} \;IL$$

$$\frac{}{\mathcal{X}_\mathcal{S}, \mathcal{Z}_\mathcal{S}; !A_j, [\varnothing; \langle\rangle\mathbf{I}], a_1, \ldots a_i, a_{i+1}, \ldots, a_n \Rightarrow s} \;\langle\rangle L$$

$$\frac{}{\mathcal{X}_\mathcal{S}, \mathcal{Z}_\mathcal{S}; !A_j, \langle\rangle\langle\rangle\mathbf{I}, a_1, \ldots a_i, a_{i+1}, \ldots, a_n \Rightarrow s} \;\langle\rangle L$$

$$\frac{}{\mathcal{X}_\mathcal{S}, \mathcal{Z}_\mathcal{S}; !A_j \bullet \langle\rangle\langle\rangle\mathbf{I}, a_1, \ldots a_i, a_{i+1}, \ldots, a_n \Rightarrow s} \;\bullet L$$

$$\frac{}{\mathcal{X}_\mathcal{S}, \mathcal{Z}_\mathcal{S}; [\varnothing; []^{-1}(!A_j \bullet \langle\rangle\langle\rangle\mathbf{I})], a_1, \ldots, a_i, a_{i+1}, \ldots, a_n \Rightarrow s} \;[]^{-1}L$$

$$\frac{}{\mathcal{X}_\mathcal{S}, \mathcal{Z}_\mathcal{S}; [\varnothing; ![]^{-1}(!A_j \bullet \langle\rangle\langle\rangle\mathbf{I})], a_1, \ldots, a_i, a_{i+1}, \ldots, a_n \Rightarrow s} \;!D$$

$$\frac{}{\mathcal{X}_\mathcal{S}, \mathcal{Z}_\mathcal{S}; [![]^{-1}(!A_j \bullet \langle\rangle\langle\rangle\mathbf{I}); \Lambda], a_1, \ldots, a_i, a_{i+1}, \ldots, a_n \Rightarrow s} \;!P$$

$$\frac{}{\mathcal{X}_\mathcal{S}, \mathcal{Z}_\mathcal{S}; [\varnothing; [\varnothing; \Lambda]], a_1, \ldots, a_i, a_{i+1}, \ldots, a_n \Rightarrow s} \;!C \;\; (![]^{-1}(!A_j \bullet \langle\rangle\langle\rangle\mathbf{I}) = !Z_i \in \mathcal{Z}_\mathcal{S})$$

Using the *land* rule, the last rewriting step, from $a_1 \ldots a_i x_1 \ldots x_m a_r \ldots a_n$ to $a_1 \ldots a_i y_1 \ldots y_k a_r \ldots a_n$ is simulated as follows. Since $x_1 \ldots x_m \to y_1 \ldots y_k$ is a rewriting rule of $\mathcal{S}$, the formula $A_j = (x_1 \bullet \ldots \bullet x_m)/(y_1 \bullet \ldots \bullet y_k)$ belongs to $\mathcal{A}_\mathcal{S}$. Thus, the *land* rule is applicable.

$$\frac{\dfrac{\varnothing; y_1 \Rightarrow y_1 \quad \ldots \quad \varnothing; y_k \Rightarrow y_k}{\varnothing; y_1, \ldots, y_k \Rightarrow y_1 \bullet \ldots \bullet y_k}\;\bullet R \quad \dfrac{\dfrac{\mathcal{X}_\mathcal{S}, \mathcal{Z}_\mathcal{S}; [\varnothing; [\varnothing; \Lambda]], a_1, \ldots, a_i, x_1, \ldots, x_m, a_r, \ldots, a_n \Rightarrow s}{\mathcal{X}_\mathcal{S}, \mathcal{Z}_\mathcal{S}; [\varnothing; [\varnothing; \Lambda]], a_1, \ldots, a_i, x_1 \bullet \ldots \bullet x_m, a_r, \ldots, a_n \Rightarrow s}\;\bullet L}{\mathcal{X}_\mathcal{S}, \mathcal{Z}_\mathcal{S}; [\varnothing; [\varnothing; \Lambda]], a_1, \ldots, a_i, (x_1 \bullet \ldots \bullet x_m)/(y_1 \bullet \ldots \bullet y_k), y_1, \ldots, y_k, a_r, \ldots, a_n \Rightarrow s}\;/L}{\mathcal{X}_\mathcal{S}, \mathcal{Z}_\mathcal{S}; [\varnothing; [\varnothing; \Lambda]], a_1, \ldots, a_i, y_1, \ldots, y_k, a_r, \ldots, a_n \Rightarrow s}\;land$$

For the case of empty $x_1 \ldots x_m$ or $y_1 \ldots y_m$ the derivations are a bit different:

$$\frac{\dfrac{\varnothing; y_1 \Rightarrow y_1 \quad \ldots \quad \varnothing; y_k \Rightarrow y_k}{\varnothing; y_1, \ldots, y_k \Rightarrow y_1 \bullet \ldots \bullet y_k}\;\bullet R \quad \dfrac{\dfrac{\mathcal{X}_\mathcal{S}, \mathcal{Z}_\mathcal{S}; [\varnothing; [\varnothing; \Lambda]], a_1, \ldots, a_i, a_r, \ldots, a_n \Rightarrow s}{\mathcal{X}_\mathcal{S}, \mathcal{Z}_\mathcal{S}; [\varnothing; [\varnothing; \Lambda]], a_1, \ldots, a_i, \mathbf{I}, a_r, \ldots, a_n \Rightarrow s}\;IL}{\mathcal{X}_\mathcal{S}, \mathcal{Z}_\mathcal{S}; [\varnothing; [\varnothing; \Lambda]], a_1, \ldots, a_i, \mathbf{I}/(y_1 \bullet \ldots \bullet y_k), y_1, \ldots, y_k, a_r, \ldots, a_n \Rightarrow s}\;/L}{\mathcal{X}_\mathcal{S}, \mathcal{Z}_\mathcal{S}; [\varnothing; [\varnothing; \Lambda]], a_1, \ldots, a_i, y_1, \ldots, y_k, a_r, \ldots, a_n \Rightarrow s}\;land$$

$$\frac{\dfrac{}{\varnothing; \Lambda \Rightarrow \mathbf{I}}\;IR \quad \dfrac{\dfrac{\mathcal{X}_\mathcal{S}, \mathcal{Z}_\mathcal{S}; [\varnothing; [\varnothing; \Lambda]], a_1, \ldots, a_i, x_1, \ldots, x_m, a_r, \ldots, a_n \Rightarrow s}{\mathcal{X}_\mathcal{S}, \mathcal{Z}_\mathcal{S}; [\varnothing; [\varnothing; \Lambda]], a_1, \ldots, a_i, x_1 \bullet \ldots \bullet x_m, a_r, \ldots, a_n \Rightarrow s}\;\bullet L}{\mathcal{X}_\mathcal{S}, \mathcal{Z}_\mathcal{S}; [\varnothing; [\varnothing; \Lambda]], a_1, \ldots, a_i, (x_1 \bullet \ldots \bullet x_m)/\mathbf{I}, a_r, \ldots, a_n \Rightarrow s}\;/L}{\mathcal{X}_\mathcal{S}, \mathcal{Z}_\mathcal{S}; [\varnothing; [\varnothing; \Lambda]], a_1, \ldots, a_i, a_r, \ldots, a_n \Rightarrow s}\;land$$

$\boxed{2 \Rightarrow 3}$ By Proposition 2, since $\mathcal{X}_\mathcal{S}, \mathcal{Z}_\mathcal{S}; [\varnothing; [\varnothing; \Lambda]], a_1, \ldots, a_n \Rightarrow s$ is derivable in $!^\mathbf{s}_{\mathbf{b}1/2}\mathbf{L}^*\mathbf{b}$, $\pi(\mathcal{X}_\mathcal{S}, \mathcal{Z}_\mathcal{S}; [\varnothing; [\varnothing; \Lambda]], a_1, \ldots, a_n) \Rightarrow s$ is derivable in $!\mathbf{L}^*$. The π-projection of Z_i is $!A_i \bullet \mathbf{I}$ and the π-projection of $\mathbf{I}/(\langle\rangle\langle\rangle\mathbf{I})$ is $\mathbf{I}/\mathbf{I}$. Thus, by definition of π,

$$\pi(\mathcal{X}_\mathcal{S}, \mathcal{Z}_\mathcal{S}; [\varnothing; [\varnothing; \Lambda]], a_1, \ldots, a_n) =$$
$$!(\mathbf{I}/!(!A_1 \bullet \mathbf{I})), \ldots, !(\mathbf{I}/!(!A_n \bullet \mathbf{I})), !(\mathbf{I}/\mathbf{I}), !!(!A_1 \bullet \mathbf{I}), \ldots, !!(!A_n \bullet \mathbf{I}), a_1, \ldots, a_n,$$

and the sequent

$$!(\mathbf{I}\,/\,!(!A_1 \bullet \mathbf{I})), \ldots, !(\mathbf{I}\,/\,!(!A_n \bullet \mathbf{I})), !(\mathbf{I}\,/\,\mathbf{I}), !!(!A_1 \bullet \mathbf{I}), \ldots, !!(!A_n \bullet \mathbf{I}), a_1, \ldots, a_n \Rightarrow s$$

is derivable in $!\mathbf{L}^*$. Next, $\Lambda \Rightarrow !(\mathbf{I}\,/\,!(!A_i \bullet \mathbf{I}))$, $\Lambda \Rightarrow !(\mathbf{I}\,/\,\mathbf{I})$, and $!A_i \Rightarrow !!(!A_i \bullet \mathbf{I})$ are derivable in $!\mathbf{L}^*$:

$$\cfrac{\cfrac{\cfrac{\Lambda \Rightarrow \mathbf{I}}{!(!A_i \bullet \mathbf{I}) \Rightarrow \mathbf{I}}\,!W}{\Lambda \Rightarrow \mathbf{I}\,/\,!(!A_i \bullet \mathbf{I})}\,/R}{\Lambda \Rightarrow !(\mathbf{I}\,/\,!(!A_i \bullet \mathbf{I}))}\,!R
\qquad
\cfrac{\cfrac{\mathbf{I} \Rightarrow \mathbf{I}}{\Lambda \Rightarrow \mathbf{I}\,/\,\mathbf{I}}\,/R}{\Lambda \Rightarrow !(\mathbf{I}\,/\,\mathbf{I})}\,!R
\qquad
\cfrac{\cfrac{\cfrac{!A_i \Rightarrow !A_i \quad \Lambda \Rightarrow \mathbf{I}}{!A_i \Rightarrow !A_i \bullet \mathbf{I}}\,\bullet R}{!A_i \Rightarrow !(!A_i \bullet \mathbf{I})}\,!R}{!A_i \Rightarrow !!(!A_i \bullet \mathbf{I})}\,!R$$

Using cut in $!\mathbf{L}^*$, we obtain

$$!A_1, \ldots, !A_n, a_1, \ldots, a_n \Rightarrow s,$$

which is exactly the necessary $\Gamma_{\mathcal{S}}, a_1, \ldots, a_n \Rightarrow s$.

Next, we can eliminate applications of cut in the $!\mathbf{L}^*$ derivation by Proposition 1.

$\boxed{3 \Rightarrow 1}$ This part comes directly from the standard undecidability proof for $!\mathbf{L}^*$, see [16]. Consider the derivation of $\Gamma_{\mathcal{S}}, a_1, \ldots, a_n \Rightarrow s$ in $!\mathbf{L}^*$. Recall that the cut rule can be eliminated by Proposition 1, so we can suppose that this derivation is cut-free. All formulae in this derivation are subformulae of the goal sequent, and the only applicable rules are $\bullet L$, $\bullet R$, $/L$, and rules operating ! in the antecedent: $!L$, $!C_{1,2}$, $!W$.

Now let us hide all the formulae which include $/$. Since all formulae with ! in our sequent included $/$, this trivialises all !-operating rules. Next, let us replace all $\bullet$'s in the antecedents with commas, and remove unnecessary $\mathbf{I}$'s there. This, in its turn, trivialises $\bullet L$ and $\mathbf{IL}$. All sequents in our derivation are now of the form $b_1, \ldots, b_s \Rightarrow C$, where $s \geq 0$ and $C = c_1 \bullet \ldots \bullet c_r$ $(r \geq 1)$ or $C = \mathbf{I}$. For the sake of uniformity, we also write $C = \mathbf{I}$ as $C = c_1 \bullet \ldots \bullet c_r$ with $r = 0$. Inference rules reduce to

$$\cfrac{b_{i+1}, \ldots, b_j \Rightarrow y_1 \bullet \ldots \bullet y_k \quad b_1, \ldots, b_i, x_1, \ldots, x_m, b_{j+1}, \ldots, b_s \Rightarrow C}{b_1, \ldots, b_i, b_{i+1}, \ldots, b_j, b_{j+1}, \ldots, b_s \Rightarrow C}$$

where $x_1, \ldots, x_m \to y_1, \ldots y_k$ is a rewriting rule of $\mathcal{S}$;

$$\cfrac{b_1, \ldots, b_i \Rightarrow c_1 \bullet \ldots \bullet c_j \quad b_{i+1}, \ldots, b_s \Rightarrow c_{j+1} \bullet \ldots \bullet c_r}{b_1, \ldots, b_i, b_{i+1}, \ldots, b_s \Rightarrow c_1 \bullet \ldots \bullet c_j \bullet c_{j+1} \bullet \ldots \bullet c_r}$$

and, finally, we have axioms of the form $a \Rightarrow a$ and $\Lambda \Rightarrow \mathbf{I}$.

Now straightforward induction on derivation establishes the following fact: if $b_1, \ldots, b_s \Rightarrow c_1 \bullet \ldots \bullet c_r$ is derivable in the simplified calculus presented above, then $b_1 \ldots b_s$ is derivable from $c_1 \ldots c_r$ in the semi-Thue system $\mathcal{S}$. This finishes our proof.

5   Conclusion

In this paper, we have discussed a new version of interaction between brackets and exponential, recently proposed by Glyn Morrill [31,32]. This system is intended to be the basis for the categorial grammar parser CatLog3. For a fragment of this system, we have proved undecidability of the derivability problem. Undecidability for the corresponding fragment of a previous version [28] of Morrill's system was shown in [14]. The new contraction rule introduced by Morrill, however, significantly differs from the earlier ones, and, unfortunately, existing undecidability proofs [12,14,23] do not directly extend to the new version. The necessary new technique for proving undecidability with the new form of the contraction rule [31,32] was developed in the present paper.

Future Work

One of the referees pointed out the following interesting question. The calculus $!_{b1/2}^s L^* b$, considered in this paper, can generate ungrammatical sentences (see end of Sect. 2), since it allows the user to put brackets on empty substrings of the sentence being parsed. The question is whether the undecidability proof presented in this paper is still valid for the variant of $!_{b1/2}^s L^* b$ where such bracketing is disallowed. Furthermore, for the sake of cut-elimination, this non-emptiness restriction should possibly be propagated to all bracketed expressions and generally all meta-formulae inside the derivation. In particular, this condition would require excluding the product unit, $\mathbf{I}$. The product unit is essentially used in our undecidability proof, but potentially could be replaced by a unit-free formula (cf. [19]). We leave this problem open for future research. There are also issues with reconciling non-emptiness restrictions, cut-elimination, the substitution property, and the full-power exponential modality [11,13]. Settling these issues for $!_{b1/2}^s L^* b$, the calculus with brackets and non-standard rules for !, requires further investigation.

There are several other problems which are still open. One open problem is whether syntactic condition could be imposed on the formulae under ! (like the so-called bracket non-negative condition [14,28]), under which the system becomes decidable. There is also an issue of extending the bracket-inducing algorithm from [30] to the system with the subexponential discussed in the present paper. Finally, it is interesting whether our result could be strengthened to the undecidability of the one-division fragment of $!_{b1/2}^s L^* b$, as it was done in [12] using Buszkowski's technique [4] of encoding semi-Thue derivations in the one-divison Lambek calculus.

Acknowledgments. The authors are thankful to the anonymous referees for helpful comments and interesting questions. They would also like to thank Glyn Morrill for fruitful discussions of linguistic motivations for the calculi Morrill introduced in [27–29,31,32], which are considered in this paper.

Financial Support
The work of Max Kanovich and Andre Scedrov was supported by the Russian Science Foundation under grant 17-11-01294 and performed at National Research University Higher School of Economics, Moscow, Russia. The work of Stepan Kuznetsov was supported by the Young Russian Mathematics award, by the grant MK-430.2019.1 of the President of Russia, and by the Russian Foundation for Basic Research grant 18-01-00822. Section 3 was contributed by Kanovich and Scedrov. Section 4 was contributed by Kuznetsov. Sections 1, 2, and 5 were contributed jointly and equally by all co-authors.

References

1. Abrusci, V.M.: A comparison between Lambek syntactic calculus and intuitionistic linear logic. Zeitschr. Math. Logik Grundl. Math. **36**, 11–15 (1990). https://doi.org/10.1002/malq.19900360103
2. Ajdukiewicz, K.: Die syntaktische Konnexität. Stud. Philos. **1**, 1–27 (1935)
3. Bar-Hillel, Y.: A quasi-arithmetical notation for syntactic description. Language **29**, 47–58 (1953)
4. Buszkowski, W.: Some decision problems in the theory of syntactic categories. Zeitschr. Math. Logik Grundl. Math. **28**, 539–548 (1982). https://doi.org/10.1002/malq.19820283308
5. Buszkowski, W., Palka, E.: Infinitary action logic: complexity models and grammars. Stud. Logica. **89**(1), 1–18 (2008)
6. Chomsky, N.: Three models for the description of language. IRE Trans. Inf. Theory I T–**2**(3), 113–124 (1956)
7. Girard, J.-Y.: Linear logic. Theor. Comput. Sci. **50**(1), 1–102 (1987). https://doi.org/10.1016/0304-3975(87)90045-4
8. Girard, J.-Y.: A new constructive logic: classical logic. Math. Struct. Comput. Sci. **1**(3), 255–296 (1991). https://doi.org/10.1017/S0960129500001328
9. Girard, J.-Y.: On the unity of logic. Ann. Pure Appl. Logic **59**(3), 201–217 (1993). https://doi.org/10.1016/0168-0072(93)90093-S
10. Girard, J.-Y.: Light linear logic. Inf. Comput. **143**(2), 175–204 (1998). https://doi.org/10.1006/inco.1998.2700
11. Kanovich, M., Kuznetsov, S., Scedrov, A.: On Lambek's restriction in the presence of exponential modalities. In: Artemov, S., Nerode, A. (eds.) LFCS 2016. LNCS, vol. 9537, pp. 146–158. Springer, Cham (2016). https://doi.org/10.1007/978-3-319-27683-0_11
12. Kanovich, M., Kuznetsov, S., Scedrov, A.: Undecidability of the Lambek calculus with a relevant modality. In: Foret, A., Morrill, G., Muskens, R., Osswald, R., Pogodalla, S. (eds.) FG 2015-2016. LNCS, vol. 9804, pp. 240–256. Springer, Heidelberg (2016). https://doi.org/10.1007/978-3-662-53042-9_14
13. Kanovich, M., Kuznetsov, S., Scedrov, A.: Reconciling Lambek's restriction, cut-elimination, and substitution in the presence of exponential modalities. Annals Pure Applied Logic, accepted for publication. arXiv:1608.02254 (2016)
14. Kanovich, M., Kuznetsov, S., Scedrov, A.: Undecidability of the Lambek calculus with subexponential and bracket modalities. In: Klasing, R., Zeitoun, M. (eds.) FCT 2017. LNCS, vol. 10472, pp. 326–340. Springer, Heidelberg (2017). https://doi.org/10.1007/978-3-662-55751-8_26

15. Kanovich, M., Kuznetsov, S., Scedrov, A.: Lambek calculus enriched with multiplexing (abstract). In: International Conference of Mal'tsev Meeting 2018, Collection of Abstracts. Sobolev Institute of Mathematics and Novosibirsk State University, Novosibirsk (2018). http://www.math.nsc.ru/conference/malmeet/18/maltsev18.pdf

16. Kanovich, M., Kuznetsov, S., Nigam, V., Scedrov, A.: Subexponentials in noncommutative linear logic. Math. Struct. Comput. Sci. (2018). https://doi.org/10.1017/S0960129518000117. Accessed 2 May 2018

17. Kanovich, M., Kuznetsov, S., Nigam, V., Scedrov, A.: A logical framework with commutative and non-commutative subexponentials. In: Galmiche, D., Schulz, S., Sebastiani, R. (eds.) IJCAR 2018. LNCS (LNAI), vol. 10900, pp. 228–245. Springer, Cham (2018). https://doi.org/10.1007/978-3-319-94205-6_16

18. Kozen, D.: On the complexity of reasoning in Kleene algebra. Inf. Comput. **179**, 152–162 (2002). https://doi.org/10.1006/inco.2001.2960

19. Kuznetsov, S.L.: On the Lambek calculus with a unit and one division. Moscow Univ. Math. Bull. **66**(4), 173–175 (2011). https://doi.org/10.3103/S0027132211040085

20. Kuznetsov, S.: The Lambek calculus with iteration: two variants. In: Kennedy, J., de Queiroz, R.J.G.B. (eds.) WoLLIC 2017. LNCS, vol. 10388, pp. 182–198. Springer, Heidelberg (2017). https://doi.org/10.1007/978-3-662-55386-2_13

21. Lafont, Y.: Soft linear logic and polynomial time. Theor. Comput. Sci. **318**(12), 163–180 (2004). https://doi.org/10.1016/j.tcs.2003.10.018

22. Lambek, J.: The mathematics of sentence structure. Am. Math. Mon. **65**, 154–170 (1958). https://doi.org/10.2307/2310058

23. Lincoln, P., Mitchell, J., Scedrov, A., Shankar, N.: Decision problems for propositional linear logic. Ann. Pure Appl. Logic **56**(1–3), 239–311 (1992). https://doi.org/10.1016/0168-0072(92)90075-B

24. Markov, A.: On the impossibility of certain algorithms in the theory of associative systems. Doklady Acad. Sci. USSR (N. S.) **55**, 583–586 (1947)

25. Moortgat, M.: Multimodal linguistic inference. J. Logic Lang. Inf. **5**(3/4), 349–385 (1996). https://doi.org/10.1007/BF00159344

26. Morrill, G.: Categorial formalisation of relativisation: pied piping, islands, and extraction sites. Technical report LSI-92-23-R, Universitat Politècnica de Catalunya (1992)

27. Morrill, G.V.: Categorial Grammar: Logical Syntax, Semantics, and Processing. Oxford University Press, Oxford (2011)

28. Morrill, G., Valentín, O.: Computational coverage of TLG: nonlinearity. In: Proceedings of NLCS 2015. EPiC Series, vol. 32, pp. 51–63 (2015)

29. Morrill, G.: Grammar logicised: relativisation. Linguist. Philos. **40**(2), 119–163 (2017). https://doi.org/10.1007/s10988-016-9197-0

30. Morrill, G., Kuznetsov, S., Kanovich, M., Scedrov, A.: Bracket induction for Lambek calculus with bracket modalities. In: Foret, A., Kobele, G., Pogodalla, S. (eds.) FG 2018. LNCS, vol. 10950, pp. 84–101. Springer, Heidelberg (2018). https://doi.org/10.1007/978-3-662-57784-4_5

31. Morrill, G.: A note on movement in logical grammar. J. Lang. Model. **6**(2), 353–363 (2018). https://doi.org/10.15398/jlm.v6i2.233

32. Morrill, G.: Parsing/theorem-proving for logical grammar CatLog3. J. Logic Lang. Inf. (2019). https://doi.org/10.1007/s10849-018-09277-w. Accessed 18 Jan 2019

33. Post, E.L.: Recursive unsolvability of a problem of Thue. J. Symb. Logic **12**, 1–11 (1947)

34. Thue, A.: Probleme über Veränderungen von Zeichenreihen nach gegebenen Regeln. Kra. Vidensk. Selsk. Skrifter. **10**, 1–34 (1914)

Proof-Theoretic Aspects of Hybrid Type-Logical Grammars

Richard Moot[1(✉)] and Symon Jory Stevens-Guille[2]

[1] LIRMM, Université de Montpellier, CNRS, Montpellier, France
Richard.Moot@lirmm.fr
[2] Ohio State University, Columbus, USA

Abstract. This paper explores proof-theoretic aspects of hybrid type-logical grammars, a logic combining Lambek grammars with lambda grammars. We prove some basic properties of the calculus, such as normalisation and the subformula property and also present a proof net calculus for hybrid type-logical grammars. In addition to clarifying the logical foundations of hybrid type-logical grammars, the current study opens the way to variants and extensions of the original system, including but not limited to a non-associative version and a multimodal version incorporating structural rules and unary modes.

Keywords: Lambek calculus · Lambda grammar ·
Type-logical grammar · Proof theory · Proof nets

1 Introduction

Hybrid type-logical grammars (HTLG), a logic introduced by Kubota and Levine [7], combines the standard Lambek grammar implications with the lambda grammar operations. As a consequence, the lambda calculus term constructors of abstraction and application live side-by-side with the Lambek calculus operation of concatenation and its residuals. The logic is motivated by empirical limitations of its subsystems. It provides a simple account of many phenomena on the syntax-semantics interface, for which neither of its subsystems has equally simple solutions [5–7].

For instance, Lambek calculi struggle to account for medial extraction, as is required for the wide-scope reading of the universal in (1). Such cases are straightforwardly accounted for by lambda grammars. For the same reasons—namely the absence of directionality—lambda grammars cannot easily distinguish (2) from (3) [12], whereas the distinction is trivial to implement in Lambek calculi.

1. Someone delivers every letter to its destination.
2. *Ahmed loves and dislikes dessert the pizza
3. Ahmed loves and Johani dislikes the pizza

© Springer-Verlag GmbH Germany, part of Springer Nature 2019
R. Bernardi et al. (Eds.): FG 2019, LNCS 11668, pp. 84–100, 2019.
https://doi.org/10.1007/978-3-662-59648-7_6

In their paper on determiner gapping in hybrid type-logical grammar, Kubota and Levine [6, footnote 7] 'acknowledge that there remains an important theoretical issue: the formal properties of our hybrid implicational logic are currently unknown'. In this paper, we prove basic properties of the natural deduction calculus of Kubota and Levine [7] (normalisation, decidability and the subformula property) and present a proof net calculus. This puts HTLG on a firm theoretical foundation, but also provides a framework for extensions of the logic. In addition, the proof net calculus provides a proof search method which is both flexible and transparent.

2 Natural Deduction

HTLG syntactic terms are tuples of which the first element is a linear lambda term and the second is a type-logical formula drawn from the union of implicational linear logic and Lambek formulas. Given a set of atomic formulas A (we will assume A contains at least the atomic formula n for noun, np for noun phrase, s for sentence, and pp for prepositional phrase), the formula language of HTLG is the following.

- $T_{Logic} ::= T_{Lambek} \mid T_{Logic} \multimap T_{Logic}$
- $T_{Lambek} ::= A \mid T_{Lambek}/T_{Lambek} \mid T_{Lambek}\backslash T_{Lambek}$

Prosodic types are simple types with a unique atomic type s (for *structure* or, in an associative context, *string*). Logical formulas are translated to prosodic types as follows.

$$Pros(T_{\text{Lambek}}) = s$$
$$Pros(T_{\text{Logic}} \multimap T_{\text{Logic}}) = Pros(T_{\text{Logic}}) \to Pros(T_{\text{Logic}})$$

The lambda terms of HTLG, called *prosodic terms*, are constructed as follows.

- Atoms: $+^{s \to s \to s}$, ϵ^s, a countably infinite number of variables $x_0, x_1, \ldots$ for each type α; by convention we use $p, q, \ldots$ for variables of type s.
- Construction rules:
  - if $M^{\alpha \to \beta}$ and N^α, then $(MN)^\beta$
  - if x^α and M^β, then $(\lambda x.M)^{\alpha \to \beta}$

In what follows, we restrict the prosodic terms to linear lambda terms, requiring each λ binder to bind exactly one occurrence of its variable x. This restriction is standard in HTLG.

The natural deduction rules for HTLG are given by Fig. 1. The lexicon assigns to the word p a formula A and a linear lambda term M of type $Pros(A)$. Since this term is linear, it contains exactly one free occurrence of p. When no confusion is possible (for example when a word appears several times in a sentence), we use the word itself instead of the unique variable p (the formula w has a purely technical role and cannot appear on the right hand side of the lexicon or axiom rule). An example would be $\lambda P.(P \; everyone) : (np \multimap s) \multimap s$. The elimination

$$\frac{}{p^s : w \vdash M : A} \; \text{Lex} \qquad \frac{}{x^\alpha : A \vdash x^\alpha : A} \; \text{Ax}$$

$$\frac{\Gamma \vdash N^\alpha : A \qquad \Delta \vdash M^{\alpha \to \beta} : A \multimap B}{\Gamma, \Delta \vdash (MN)^\beta : B} \multimap E \qquad \frac{\Gamma, x^\alpha : A \vdash M^\beta : B}{\Gamma \vdash (\lambda x.M)^{\alpha \to \beta} : A \multimap B} \multimap I$$

$$\frac{\Gamma \vdash M^s : A/B \qquad \Delta \vdash N^s : B}{\Gamma, \Delta \vdash (M+N)^s : A} \; /E \qquad \frac{\Gamma, p^s : A \vdash (M+p)^s : B}{\Gamma \vdash M^s : B/A} \; /I$$

$$\frac{\Delta \vdash M^s : B \qquad \Gamma \vdash N^s : B \backslash A}{\Delta, \Gamma \vdash (M+N)^s : A} \; \backslash E \qquad \frac{p^s : A, \Gamma \vdash (p+M)^s : B}{\Gamma \vdash M^s : A \backslash B} \; \backslash I$$

$$\frac{\Gamma \vdash M[(\lambda x.N)P] : C}{\Gamma \vdash M[N[x := P]] : C} \; [\beta]$$

Fig. 1. Gentzen-Style ND inference rules for HTLG

rules have the standard condition that no free variables are shared between Γ and Δ, which ensures Γ, Δ is a valid context. The introduction rules have the standard side-condition that Γ contains at least one formula (ensuring that provable statements cannot have empty antecedents). The β rule performs on-the-fly beta reduction on the lambda terms.

Before showing normalization, we first prove a standard substitution lemma.

Lemma 1. *Let δ_1 be a proof of $\Gamma \vdash N : A$ and δ_2 a proof of $\Delta, x : A \vdash M[x] : C$ such that N and M share no free variables, then there is a proof of $\Gamma, \Delta \vdash M[N] : C$.*

Proof. We can combine the two proofs as follows, replacing the hypothesis $x : A$ of δ_2 by the proof δ_1.

$$\begin{array}{c} \vdots \; \delta_1 \\ \Gamma \vdash N : A \\ \vdots \; \delta_2[x := N] \\ \Gamma, \Delta \vdash M[N] : C \end{array}$$

Given that, by construction, M and N share no free variables, replacing x by N cannot make a rule application in δ_2 invalid.   □

Given that the w atomic formula appearing on the left-hand side of the Lex rule is by construction forbidden to appear on the right-hand side of a sequent, this means that the substitution lemma can never apply to a lexical hypothesis (since there are no proofs of the form $\Gamma \vdash N : w$).

3   Normalisation

We show that HTLG is normalizing. A normal form for an HTLG proof is defined as follows.

Definition 1. *A derivation D for HTLG is normal iff each major premiss of an elimination rule is either:*

1. an assumption
2. a conclusion of an application of an E-rule.

In general, we call a logic *normalizing* just in case there is an effective procedure for extracting normal proofs from arbitrary proofs. Based on this definition, any path in a normal proof starts with an axiom/lexicon rule, then passes through a (possibly empty) sequence of elimination rules as the major premiss, followed by a (possibly empty) sequence of introduction rules, ending either in the minor premiss of an elimination rule or in the conclusion of the proof.

To demonstrate HTLG is normalizing, we define a set of conversion rules—functions from derivations D to derivations D'—such that repeated application of the rules terminates in a normal derivation.

Figure 2 shows the conversion rules[1]. Note that, given the condition on the elimination rules, N and M cannot share free variables and that Lemma 1 therefore guarantees the reductions transform proofs into proofs. In what follows, we assume that β reduction applies on an as-needed basis, ignoring its application for simplicity of presentation.

$$
\cfrac{\vdots\ \Pi_1 \quad \cfrac{\cfrac{x^s : A}{\vdots\ \Pi_2}\quad (x+M)^s : B}{M^s : A\backslash B}\ \backslash I}{(N+M)^s : B}\ \backslash E
\qquad\rightsquigarrow\qquad
\cfrac{\vdots\ \Pi_1}{N^s : A}\ \ \cfrac{\vdots\ \Pi_2[x := N]}{(N+M)^s : B}
$$

$$
\cfrac{\vdots\ \Pi_1 \quad \cfrac{\cfrac{x^\alpha : A}{\vdots\ \Pi_2}\quad M^\beta : B}{(\lambda x.M)^{\alpha\to\beta} : A \multimap B}\ \multimap I}{((\lambda x.M)N)^\beta : B}\ \multimap E
\qquad\rightsquigarrow\qquad
\cfrac{\vdots\ \Pi_1}{N^\alpha : A}\ \ \cfrac{\vdots\ \Pi_2[x := N]}{(M[x := N])^\beta : B}
$$

Fig. 2. Conversion rules

Theorem 1. *HTLG is strongly normalizing.*

Proof. To show strong normalization, we need to show that there are no infinite reduction sequences. However, since each reduction reduces the size of the proof, this is trivial. □

Theorem 2. *Normalization for HTLG proofs is confluent.*

---

[1] The rule for the / directly parallels that for \, modulo directionality.

Proof. It is easy to show weak confluence: whenever a proof can be reduced by two different reductions R_1 and R_2, then reducing either redex will preserve the other redex , and R_1 followed by R_2 will produce the same proof as R_2 followed by R_1. By Theorem 1 we therefore have strong confluence.   □

Corollary 1. *HTLG proofs have a unique normal form.*

Proof. Immediate by Theorems 1 and 2. Note that uniqueness is up to beta equivalence or, alternatively, each HTLG proof has a unique normal form proof with a beta normal term[2].   □

Corollary 2. *HTLG satisfies the subformula property.*

Proof. This is a direct consequence of normalization (Theorem 1). In a normal form proof, every formula is either a subformula of one of the hypotheses or a subformula of the conclusion.   □

Given that we only consider linear lambda terms, HTLG proofs have a number of beta reductions bounded from above by the total number of abstractions in the proof (those in the leaves plus those in the introduction rules). Therefore, decidability follows from the subformula property. However, we will give a more detailed complexity analysis in Sect. 6.

4   Proof Nets

In proof theory, we generally have multiple proof systems for a single logic. Even though deductions in these systems are intertranslatable, shifting to a different proof system may make some properties of the logic easier to prove.

Proof nets are a graph theoretic representation of proofs introduced for linear logic by Girard [3]. Proof nets remove the possibility of 'boring' rule permutations as they occur in the sequent calculus or natural deduction[3], solving the so-called problem of 'spurious ambiguity' in type-logical grammars.

We generally define proof nets as part of a larger class called *proof structures*. Proof nets are those proof structures which correspond to sequent (or natural deduction) proofs. We can distinguish proof nets from other proof structures by means of a correctness condition. As a guiding intuition, we have the following correspondence between sequent calculus/natural deduction proofs and proof nets.

$$\text{logical rule } = \text{link} + \text{correctness condition}$$

$$\text{proof (proof net) } = \text{proof structure} + \text{correctness condition}$$

A more procedural interpretation of this is that a proof structures represent the search space for proofs.

---

[2] As is usual in the lambda calculus, we do not distinguish alpha-equivalent lambda terms.

[3] For natural deduction, rule permutations are a problem only for the $\bullet E$ and the $\Diamond E$ rules.

4.1 Proof Structures

Definition 2. *A link is tuple consisting of a type (tensor or par), an index (from a fixed alphabet I, indicating the family of connectives it belongs to), a list of premisses, a list of conclusions, and an optional main node (either one of the conclusions or one of the premisses).*

A link is essentially a labelled hyperedge connecting a number of vertices in a hypergraph. The premisses of a link are drawn left-to-right above the central node, whereas the conclusions are drawn left-to-right below the central node. A par link displays the central node as a filled circle, whereas a tensor uses an open circle. For hybrid type-logical grammars, the set of indices is $\{\epsilon, +, @, \lambda\}$. The constructor ϵ represents the empty string (it doesn't correspond to a logical connective, although we can add one if desired). The (non-associative) Lambek calculus implications ($\backslash$, $/$) use the term constructor '+' for their links (in a multimodal context we can have multiple instances of '+', for example, '$+_1$', '$+_2$', but this doesn't change much), whereas the lambda grammar implication ($-\circ$) uses links labeled with @ (representing application, for its tensor link) and λ (representing abstraction, for its par link).

Table 1. Links for HTLG proof structures

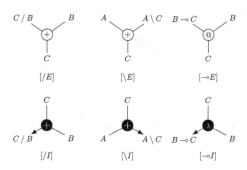

From Table 1, it is clear that par links have one premiss and two conclusions, whereas tensor links have two premisses and one conclusion (we will see a tensor link with one premiss and two conclusions later). Par links have an arrow pointing to the main formula of the link, the main formulas of tensor links are not distinguished visually (but can be determined from the formula labels).

Definition 3. *A proof structure is a tuple $\langle F, L \rangle$, where F is a set of formula occurrences (vertices labeled with formulas) and L is a set of links such that each local neighbourhood is an instance of one of the links of Table 1, and such that:*

- *each formula is at most once the premiss of a link,*
- *each formula is at most once the conclusion of a link.*

The formulas which are not a conclusion of any link in a proof structure are its hypotheses. *We distinguish between* lexical *hypotheses and* logical *hypotheses; lexical hypotheses are formulas from the lexicon, all other hypotheses are logical. The formulas which are not a premiss of any link in a proof structure are its* conclusions. *Formulas which are both a premiss and a conclusion of a link are internal nodes of the proof structure.*

We say a proof structure with hypotheses Γ and conclusions Δ is a proof structure of $\Gamma \vdash \Delta$, overloading the $\vdash$ symbol.

Definition 4. *Given a proof structure P, a formula occurrence A of P is a cut formula if it is the main formula of two links. A is an axiomatic formula in case it is not the main formula of any link.*

Example 1. As a very simple example, consider the lexicon containing only the words 'everyone' of type $(np \multimap s) \multimap s$ with prosodic term $\lambda P.(P\ everyone)$ and 'sleeps' of type $np \multimap s$ with prosodic term $\lambda z.(z + sleeps)$. Unfolding the lexical entries produces the proof structure shown in Fig. 3. We use the convention of replacing lexical hypotheses with the corresponding word, so 'everyone' represents the formula $(np \multimap s) \multimap s$ and 'sleeps' the formula $np \multimap s$. There are no cut formulas in the figure, and all atomic formulas are axiomatic.

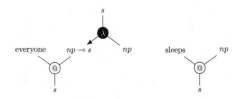

Fig. 3. Proof structure of 'everyone sleeps'.

Definition 5. *Given a proof structure P and two distinct formula occurrences x, y of P, both labeled with the same formula A, with x a logical hypothesis of P and y a conclusion of P. Then P', the vertex contraction of x and y in P, is the proof net obtained by deleting x and y, adding a new node z with label A such that z is the premiss of the link x was a premiss of (if any) and the conclusion of the link that y was the conclusion of (if any).*

The vertex contraction operation is a standard graph theoretic operation. In the current context, it operates like the cut or axiom rule in the sense that if P_1 is a proof net of $\Gamma, A \vdash \Delta$ and P_2 a proof net of Γ', A, Δ' with x and y the two occurrences of A, then the vertex contraction of x and y is a proof net of $\Gamma, \Gamma' \vdash \Delta, \Delta'$. Given that, in an intuitionistic context like the current one, all proof nets have a single conclusion we even have that if P_1 is a proof net of $\Gamma, A \vdash C$ and P_2 a proof net of $\Gamma' \vdash A$, then the vertex contraction gives a proof net of $\Gamma, \Gamma' \vdash C$. Note that vertex contraction applies only to logical hypotheses and not to lexical ones.

Just like a logical link is a generalisation of a logical rule which is locally correct but need not be correct globally, a vertex contraction is a generalisation of the cut rule which is locally correct but need not be correct globally.

Example 2. Connecting the atomic formulas of the proof structure shown in Fig. 3 produces the proof structure shown on the left of Fig. 4. It has (the formulas corresponding to) 'everyone' and 'sleeps' as hypotheses (both lexical) and the formula s as its conclusion.

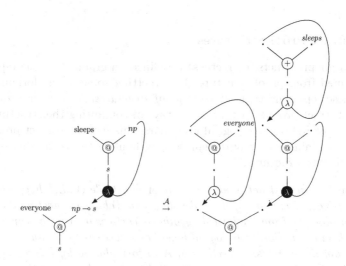

Fig. 4. Proof structure of 'everyone sleeps' after identification of the atomic formulas (left) and corresponding abstract proof structure (right).

Definition 6. *A* tensor graph *is a connected proof structure with a unique conclusion (root) node containing only tensor links. The trivial tensor graph is a single node.*

Given a proof structure P, the components *of P are the maximal substructures of P which are tensor graphs. A* tensor tree *is an acyclic tensor graph.*

For standard multimodal proof nets, we define correctness using tensor trees instead of the more general notion used here. Our results may be (graph theoretical representations of) lambda terms, and the λ link represents the λ binder for linear lambda terms. As is usual for lambda terms, we need to be careful about 'accidental capture' of variables. That is, we want avoid incorrect reductions such as $(\lambda x \lambda y (f\, x))(g\, y)$ (not a linear lambda term) to $\lambda y (f\,(g\, y))$.

Table 2. Links for HTLG abstract proof structures

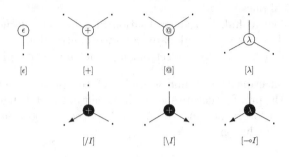

| [ε] | [+] | [@] | [λ] |
| [/I] | [\I] | [−∘I] |

4.2　Abstract Proof Structures

As is usual for proof nets, correctness is defined on graph theoretic representations obtained from proof structures by forgetting some of the formula labels. We call these representations *abstract proof structures*. A more procedural way of seeing abstract proof structures is as ways of computing the structure of the antecedent. For hybrid type-logical grammars, this means abstract proof structures must contain some way of representing lambda terms in addition to the Lambek calculus structures.

Definition 7. *An abstract proof structure A is a tuple $\langle V, L, l, h, c \rangle$ where V is a set of vertices, L is a set of the links shown in Table 2 connecting the vertices of V, l is a function from the lexical hypothesis vertices of A to the corresponding variables, h is a function from logical hypothesis vertices to formulas, and c is a function from the conclusion vertices of A to formulas (a hypothesis vertex is a vertex which is not the conclusion of any link in L, and a conclusion vertex is a vertex which is not the premiss of any link in L).*

The links for abstract proof structures are shown in Table 2. The tensor links are shown in the topmost row, the par links in the bottom row, with the par links for the Lambek connectives on the left and in the middle, and the par link for the linear implication on the bottom right.

The λ tensor link is the only non-standard link. Even though it has the same shape as the link for the Grishin connectives of Moortgat and Moot [9], it is used in a rather different way. The λ tensor link does not correspond to a logical connective but rather to lambda abstraction over terms (or rather their graph theoretical representation). To keep the logic simple and the number of connectives as small as possible, we have chosen to make the ϵ link, corresponding to the empty string, a non-logical link as well. As a consequence, ϵ can appear only in lexical terms. However, if needed, it would be easy to adapt the logic by adding a logical connective 1 corresponding to ϵ.

Definition 8. *Given a proof structure P, we obtain the corresponding abstract proof structure $\mathcal{A}(P) = A$ as follows.*

1. *we keep the set of vertices V and the set of links L of P (but we forget the formula labels of the internal nodes),*
2. *logical hypotheses are kept as simple vertices, but we replace each lexical hypothesis M : A of the proof structure by a graph g corresponding to its lambda term M, the conclusion of g is the vertex which was the lexical hypothesis of P, making the word subterm w of M a lexical hypothesis of the new structure,*
3. *we define l to assign the corresponding word for each lexical hypothesis of the resulting graph, h to assign a formula for all logical hypotheses, and c to assign a formula to all conclusions.*

Example 3. Converting the proof structure on the left of Fig. 4 to an abstract proof structure produces the abstract proof structure shown on the right. We have replaced 'everyone' by the structure corresponding to its lexical lambda term and similarly for 'sleeps'.

Definition 9. *A lambda graph is an abstract proof structure such that:*

1. *it has a single conclusion,*
2. *it contains only tensor links,*
3. *each right conclusion of a lambda link is an ancestor of its premiss,*
4. *removing the connection between all lambda links and their rightmost conclusion produces an acyclic and connected structure.*

Condition 3 avoids vacuous abstraction and accidental variable capture in the corresponding lambda term. Condition 4 is the standard acyclicity and connectedness condition for abstract proof structures, but allowing for the fact that lambda abstraction (but no other tensor links) can produce cycles.

Lambda graphs correspond to linear lambda terms in the obvious way, with the rightmost conclusion of the lambda link representing the variable abstracted over. This is a standard way of representing lambda terms in a way which avoids the necessity of variable renaming (alpha conversion).

Proposition 1. *A lambda term with free variables $x_1, \ldots, x_n$ corresponds to a lambda graph with hypotheses $x_1, \ldots, x_n$, with the @ tensor link corresponding to application, the λ tensor link to abstraction, and the + link and the ϵ link to the term constants of type $s \to s \to s$ and s respectively. To keep the terms simple, we will write $(X + Y)$ instead of $((+X)Y)$.*

4.3 Structural Rules and Contractions

To decide whether or not a given proof structure is a *proof net* (that is, corresponds to a natural deduction proof), we will introduce a system of graph rewriting. The structural rules for the non-associative version of hybrid type-logical grammars are shown on the left-hand column of Table 3. We can obtain the standard associative version simply by adding the associativity rules for '+'; more generally, we can add any structural conversion for multimodal grammars, rewriting a tensor tree into another tensor tree with the same leaves (though not necessarily in the same order) provided they do not overlap with the beta redex. The ϵ structural rules simply stipulate that 'ϵ' functions as the identity element for '$+$' (both as a left identity and as a right identity).

The key rewrite is the beta conversion rule. It is the graph theoretical equivalent of performing a beta reduction on the corresponding term. For the beta rewrite, we replace the two links (and the internal node) and perform two vertex contractions: h_1 with c_1 and h_2 with c_2. We update the functions h, l and c accordingly (if one of the h_i was in the domain of h and l then so is the resulting vertex, and similarly for the c_i and the c function of the abstract proof structure).

Table 3. Structural rules (left) and logical contractions (right) for HTLG proof nets.

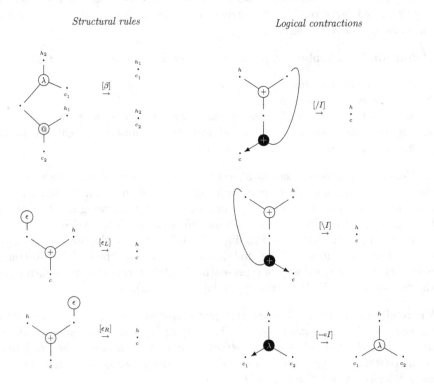

To make the operation of the beta reduction clearer, a 'sugared' version of the contraction is shown in Table 4. Term labels have been added to the vertices of the graph to make the correspondence with beta-reduction explicit. This second picture is slightly misleading in that it suggests that A_1, A_2 and A_3 are disjoint substructures. This need not be the case: for example, A_3 can contain a lambda link whose right conclusion is a premiss of either A_1 or A_2. Similarly, in a logic with the Lambek calculus product, the link for $[\bullet E]$ may connect premisses of both A_1 and A_2. A side condition on the $-\!\circ I$ conversion combined with the restriction of lexical entries to linear lambda terms will guarantee that x (c_1) in the beta reduction is always a descendant of $N(h_2)$.

Table 4. Beta conversion as a structural rule with term labels added.

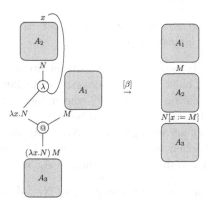

Definition 10. *We say a lambda graph is* normal *or beta-normal* *when it doesn't contain any redexes for the beta conversion.*

In addition to the structural rules, there are contractions for each of the logical connectives. Table 3 shows, on the right-hand column, the contractions for HTLG. For the Lambek implications, these are just the standard contractions. They combine a concatenation mode '+' with one of its residuals[4]. The contractions for the Lambek calculus implications are the standard contractions from Moot and Puite [11].

The contraction for $\multimap I$ has the side condition that the rightmost conclusion of the λ par link is a descendant of its premiss, passing only through tensor links. This is essentially the same condition as the one used by Danos [1], only without performing the actual contractions. This is because we want our abstract proof structures to represent the prosodic structure of a proof, which may contain lambda terms, just like the standard goal of abstract proof structures is always to compute the structure which would make the derivation valid.

Our rewrite calculus can be situated in the larger context of adding rewrite rules to the lambda calculus [2,4]. Even though the contractions for $[/I]$ and $[\backslash I]$ are not left-linear, since they correspond to terms $(M + x)/x$ and $x\backslash(x + M)$ respectively, this is not a problem because the occurrences of x are bound occurrences [4]. In general, confluence can not be maintained in the presence of structural rules (or of the unary connectives) since the structural rules themselves

---

[4] To ensure confluence of '/' and '\' in the presence of ϵ we can add the side condition to the $[/I]$ and $[\backslash I]$ contractions that the component to which the par link is attached has at least one hypothesis other than the auxiliary conclusion of the par link. This forbids empty antecedent derivations and restores confluence.

need not be confluent. Confluence of beta reduction is guaranteed by not allowing any structural rewrite to overlap with the beta redex [4].

5   Correctness of the Proof Net Calculus

Definition 11. *A proof structure is a proof net whenever its abstract proof structure converts to a lambda graph.*

The reader can easily verify that the abstract proof structure shown on the right hand side of Fig. 4 back on page 8 converts to the lambda graph *everyone+ sleeps*—after one application of the $[-\!\circ I]$ conversion and three applications of the β conversion—and is therefore a proof net.

We show that a proof net with premisses $A_1, \ldots A_k$ and conclusion C converts to a lambda graph M whenever $N_1 : A_1, \ldots, N_k : A_k \vdash M : C$ (where N_i is the lexical lambda term or variable corresponding to A_i) is derivable, and vice versa.

Lemma 2. *If δ is a natural deduction proof of $N_1 : A_1, \ldots, N_k : A_k \vdash M : C$, then we can construct a proof net with premisses $A_1, \ldots, A_n$ and conclusion C contracting to M.*

Proof. Induction on the length l of δ. If $l = 0$, then we have either an axiom $x : A$ or a lexicon rule $M : A$ (with M a linear lambda term with a single free variable w). In either case, the abstract proof structure will convert in zero steps to the required graph M.

If $l > 0$, we look at the last rule of the proof and proceed by case analysis.

If the last rule of the proof is the $/I$ rule, we are in the follow case.

$$\frac{\begin{array}{c} \vdots \ \delta \\ \Gamma, x : B \vdash N + x : A \end{array}}{\Gamma \vdash N : A/B} \, /I$$

Removing the last rule gives us the shorter proof δ, and induction hypothesis gives us a proof net of $\Gamma, x : B \vdash N + x : A$. In other words, induction hypothesis gives us a proof net of $\Gamma, B \vdash A$ such that the underlying abstract proof structure converts, using a reduction sequence ρ, to $N + x$, with x corresponding to B, as shown schematically in Fig. 5.

We need to produce a proof net of $\Gamma \vdash N : A/B$. But this is done simply by adding the $/I$ link to the proof net of the induction hypothesis and adding a final $/I$ reduction as shown in Fig. 6.

The cases for $\backslash I$ and $-\!\circ$ are similar, adding the corresponding link and conversion to the proof net obtained by induction hypothesis.

The cases for the elimination rules $/E$, $\backslash E$ and $-\!\circ E$ simply combine the two proof nets obtained by induction hypothesis with the corresponding link.

If the final rule is the β rule or a structural rule, we simply add, respectively, the β reduction and the corresponding structural conversion.   □

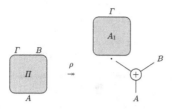

Fig. 5. Conversion sequence of Fig. 6 with the final $[/I]$ contraction removed.

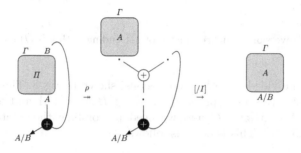

Fig. 6. Conversion sequence of a proof net ending with a $[/I]$ contraction

Lemma 3. *Given a proof net Π with premisses $A_1, \ldots A_n$ and conclusion C converting to a lambda graph M, there is a natural deduction proof $N_1 : A_1, \ldots, N_k : A_n \vdash M : C$.*

We proceed by induction on the number of conversions c.

If $c = 0$ there are no conversions. As a consequence, there are no par links in the proof net. We proceed by induction on the number of tensor links t in the proof net.

If $t = 0$, the proof net consists of a single formula A and the abstract proof structure is either a single vertex x (in the case of a hypothesis), corresponding to a proof $x : A$ or a term M corresponding to a lexical entry, corresponding to a proof $M : A$.

If $t > 0$, one of the hypotheses of the proof structure must be the main formula of its link. Removing this link will produce two proof nets and we can apply the induction hypothesis to obtain natural deduction proofs for each, combining them with the appropriate elimination rule.

If $c > 0$, we look at the last conversion and proceed by cases analysis.

If the last conversion is a β conversion or a structural rule, then induction hypothesis gives us a proof δ of $\Gamma \vdash M' : C$, which we can extend using the β rule (or structural rule) on M' to produce M and a proof of $\Gamma \vdash M : C$.

If the last conversion is a $\multimap I$ contraction, we are in the case shown in Fig. 7. The final lambda graph corresponds to the linear lambda term $M[\lambda x.N]$.

Removing the final $[\multimap I]$ conversion produces two proof nets Π_1 and Π_2 with strictly shorter conversion sequences (again because we removed the final

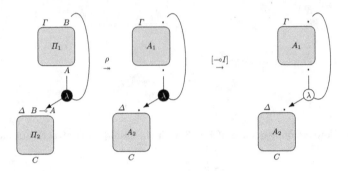

Fig. 7. Conversion sequence of a proof net ending with a $[-\!\circ I]$ contraction

conversion and divided the other conversions) shown in Fig. 8. Therefore, induction hypothesis gives us a proof δ_1 of $\Gamma, x : B \vdash N : A$ and a proof δ_2 of $\Delta, z : B \multimap A \vdash M[z] : C$ and we need to combine these into a proof of $\Gamma, \Delta \vdash M[\lambda x.N] : C$. This is done as follows.

Fig. 8. Conversion sequence of Fig. 7 with the final $[-\!\circ I]$ conversion removed

$$
\begin{array}{c}
\Gamma \quad x : B \\
\vdots \ \delta_1 \\
\dfrac{N : A}{\Delta \quad \lambda x.N : B \multimap A} \ {-\!\circ I} \\
\vdots \ \delta_2, z := \lambda x.N \\
M[\lambda x.N] : C
\end{array}
$$

The substitution of $\lambda x.N$ for z presupposes that the formula $B \multimap A$ is not a lexical hypothesis, but given that it is the conclusion of a link in the full structure, it cannot be a hypothesis of the full structure and therefore we can apply the substitution. This is the reason we distinguish between a par link for λ (which corresponds to the introduction rule of the logical connective $\multimap$) and a tensor link (which corresponds to abstraction at the term level and need not correspond to a logical rule since it can come from a complex lexical entry as well).

The cases for $/I$ and $\backslash I$ are similar, and simply follow Moot and Puite [11].

□

6 Complexity

Given the proof net calculus described in the previous sections, complexity analysis of hybrid type-logical grammars and several of its variants becomes simple.

Theorem 3. *HTLG parsing is NP complete.*

Proof. Since HTLG contains lexicalized ACG as a fragment, NP-hardness follows from Proposition 5 of Yoshinaka and Kanazawa [13], so all that remains to be shown is that HTLG is in NP.

In order to shown that HTLG parsing is NP-complete we only need to show that, given a non-deterministic proof search procedure we can verify whether a proof candidate is an actual proof in polynomial time. Given a sentence, we non-deterministically select a lexical formula for each word, then non-deterministically enumerate all proof structures for these lexical formulas. □

The proof of Theorem 3 is very general and can easily be adapted to variants and extensions of HTLG. For example, we can add the connectives for '•', '◇' and '□' and mode information (as in the multimodal versions of the Lambek calculus [8]) while maintaining NP-completeness.

When adding structural rules, complexity analysis becomes more delicate. Adding associativity, as in the original formulation of hybrid type-logical grammars, doesn't change the complexity, since we can simply use the strategy of Moot and Puite [11, Sect. 7] to ensure polynomial contraction of proof structures. So we can actually strengthen Theorem 3 to the following.

Theorem 4. $HTLG_{/_i, \bullet_i, \backslash_i, \diamond_i, \square_i}$ *parsing, with associativity for some modes i, is NP complete.*

In general, NP completeness will be preserved whenever we provide the set of structural rules with a polynomial time contraction algorithm. When we do not have a polynomial contraction algorithm, we can still show information about the complexity class: when we add structural rules but use the standard restriction that the tree rewrites allowed by the structural rules are linear (no copying or deletion of leaves) and do not increase the size of the tree, then the resulting logic is PSPACE complete, following the argument of Moot [10, Sect. 9.2].

Theorem 5. $HTLG_{/_i, \bullet_i, \backslash_i, \diamond_i, \square_i}$ *parsing with any finite set of non-expanding structural rules is PSPACE complete.*

This gives an NP lower bound and a PSPACE upper bound for any HTLG augmented with the multimodal connectives and a fixed set of structural rules, and NP completeness can be shown by providing a polynomial contraction algorithm.

7 Conclusion

We have investigated the formal properties of hybrid type-logical grammars and proved several standard results for them. This solves the question of the theoretical foundations of the system, left open by Kubota and Levine [6].

References

1. Danos, V.: La Logique Linéaire Appliquée à l'étude de Divers Processus de Normalisation (Principalement du λ-Calcul). University of Paris VII (1990)
2. Di Cosmo, R., Kesner, D.: Combining algebraic rewriting, extensional lambda calculi, and fixpoints. Theor. Comput. Sci. **169**(2), 201–220 (1996)
3. Girard, J.-Y.: Linear logic. Theor. Comput. Sci. **50**(1), 1–102 (1987)
4. Klop, J., van Oostrom, V., van Raamsdonk, F.: Combinatory reduction systems: introduction and survey. Theor. Comput. Sci. **121**(1–2), 279–308 (1993)
5. Kubota, Y., Levine, R.: Coordination in hybrid type-logical grammar. In: Ohio State University Working Papers in Linguistics, Columbus, Ohio (2013)
6. Kubota, Y., Levine, R.: Determiner gapping as higher-order discontinuous constituency. In: Morrill, G., Nederhof, M.-J. (eds.) FG 2012-2013. LNCS, vol. 8036, pp. 225–241. Springer, Heidelberg (2013). https://doi.org/10.1007/978-3-642-39998-5_14
7. Kubota, Y., Levine, R.: Gapping as like-category coordination. In: Béchet, D., Dikovsky, A. (eds.) LACL 2012. LNCS, vol. 7351, pp. 135–150. Springer, Heidelberg (2012). https://doi.org/10.1007/978-3-642-31262-5_9
8. Moortgat, M.: Categorial type logics. In: van Benthem, J., ter Meulen, A. (eds.) Handbook of Logic and Language, pp. 93–177. Elsevier/MIT Press, Amsterdam/Cambridge (1997)
9. Moortgat, M., Moot, R.: Proof nets for the Lambek-Grishin calculus. In: Grefenstette, E., Heunen, C., Sadrzadeh, M. (eds.) Quantum Physics and Linguistics: A Compositional, Diagrammatic Discourse, pp. 283–320. Oxford University Press, Oxford (2013)
10. Moot, R.: Proof nets for linguistic analysis. Utrecht Institute of Linguistics OTS, Utrecht University (2002)
11. Moot, R., Puite, Q.: Proof nets for the multimodal Lambek calculus. Stud. Logica **71**(3), 415–442 (2002)
12. Worth, C.: The phenogrammar of coordination. In: Proceedings of the EACL 2014 Workshop on Type Theory and Natural Language Semantics (TTNLS), pp. 28–36 (2014)
13. Yoshinaka, R., Kanazawa, M.: The complexity and generative capacity of lexicalized abstract categorial grammars. In: Blache, P., Stabler, E., Busquets, J., Moot, R. (eds.) LACL 2005. LNCS (LNAI), vol. 3492, pp. 330–346. Springer, Heidelberg (2005). https://doi.org/10.1007/11422532_22

On the Computational Complexity
of Head Movement and Affix Hopping

Miloš Stanojević[(✉)]

School of Informatics, University of Edinburgh,
11 Crichton Street, Edinburgh, UK
`m.stanojevic@ed.ac.uk`

Abstract. Head movement is a syntactic operation used in most generative syntactic analyses. However, its computational properties have not been extensively studied. [27] formalises head movement in the framework of Minimalist Grammars by extending the item representation to allow for easy extraction of the head. This work shows that Stabler's representation is in fact suboptimal because it causes higher polynomial parsing complexity. A new algorithm is derived for parsing head movement and affix hopping by changing the kinds of representations that the parser deals with. This algorithm has much better asymptotic worst-case runtime of $\mathcal{O}(n^{2k+5})$. This result makes parsing head movement and affix hopping computationally as efficient as parsing a single phrase movement.

Keywords: Minimalist Grammars · Parsing · Head movement · Affix hopping

1 Introduction

Minimalist Grammars (MG) [26] are a formalisation of Chomsky's Minimalist Program [4]. MGs rely on only two basic operations MERGE and MOVE. MERGE is a binary function that combines two constituents into a single constituent, while MOVE is a unary operation that takes one sub-constituent and reattaches it to the specifier position at the root of the partially constructed tree. An example MG derivation with MERGE and MOVE is shown in Fig. 1a for the declarative sentence "*[d]* she will meet him" where *[d]* is a null declarative complementiser. Figure 1b shows the X-bar structure that is a byproduct of the MG derivation and can be computed deterministically. Here MERGE combines constituents that are not necessarily adjacent to each other, while MOVE raises the subject DP from spec-VP to spec-TP so that it can check the nominative case feature.

MERGE is a single function but it is often easier to view it as different sub-functions over non-overlapping domains. For instance, merge1 is applied in the case of complement attachment, while merge2 attaches a specifier. The same holds for MOVE: move1 moves a phrase to its final landing site, while move2 moves a constituent that is going to move again later. The type of movement

© Springer-Verlag GmbH Germany, part of Springer Nature 2019
R. Bernardi et al. (Eds.): FG 2019, LNCS 11668, pp. 101–116, 2019.
https://doi.org/10.1007/978-3-662-59648-7_7

done by move1 and move2 is often called phrase movement because it is applied to a maximal projection (XP). However, phrasal movement is not the only type of movement used in Minimalist syntax. In addition to phrasal movement, all minimalist textbooks [1,2,22,25] also discuss head movement. Head movement can be triggered when a selecting lexical head merges with its complement. This operation extracts the head of the complement and adjoins it to the selecting head. It can adjoin the complement's head to the left or right of the selecting head depending on the type of feature that triggers the head movement.

The most typical example of head movement in English is Subject-Auxiliary inversion in yes-no questions. Figure 2 shows the previous example sentence turned into a question by using a different null complementiser *[q]* for forming yes-no questions. What is different between *[d]* and *[q]* is the type of the feature that each uses to select the tense phrase: *[d]* uses a simple selector =t while *[q]* uses =>t which, in addition to selecting the tense phrase, also extracts the tense head "will" and adjoins it to the left of the complementiser head.

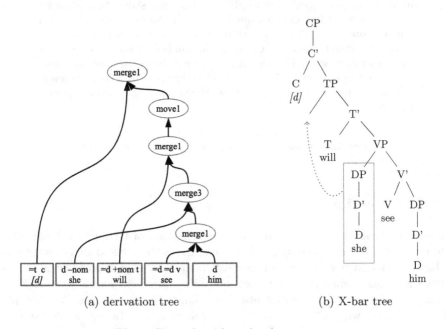

(a) derivation tree   (b) X-bar tree

Fig. 1. Example without head movement

The original version of MG published in [26] had both phrase and head movement. There have since been many variations on MG proposed, some of which are reviewed in [29] including a simpler version that does not include head movement. This simple version of MG is very convenient for formal analysis and parsing. We will call this version *succinct MG* or MG^S, and the original version of MG with head movement MG^H.

Many phenomena that appear to require head movement can in fact be expressed with other means like remnant or rightward movement [18,28]. Also,

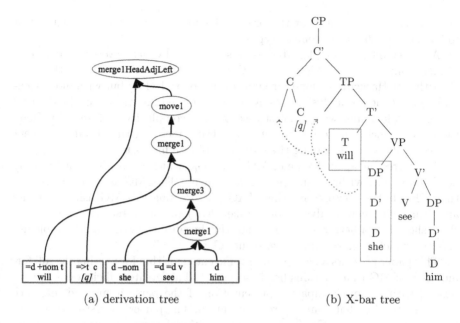

(a) derivation tree   (b) X-bar tree

Fig. 2. Example with head movement

MG with and MG without head movement are weakly equivalent [21,27]. Still, if we are interested in the structures that explain the derived word order using head movement we will need a parser specifically tailored to parsing with head movement. The practical need for that kind of parser is even more evident with the recent construction of MGbank [34] and the first wide-coverage MG parser [36], which both use head movement in almost all of their derivations.

The published work on Minimalist Parsing can be divided into two categories. The first is transition based parsing which is usually of higher relevance to psycho-linguists as a more likely model of human sentence processing [8,9,12,13,17,30–33]. However, these models use no dynamic programming and therefore have exponential runtime complexity. This makes them impractical for actual parsing and inappropriate for studying the theoretical computational complexity properties of MGs.

The second type of MG parsers are those that use dynamic programming and are usually expressed as deductive systems. These parsers run in polynomial time and are guaranteed to find a parse if it exists. The first work of that sort is Harkema's CKY parser [10,11] for MG without head movement. Stabler [27] showed a simple extension of Harkema's parser that enabled it to account for head movement. Harkema's parser for MG^S had computational time complexity of $\mathcal{O}(n^{4k+4})$ where k is a number of distinct phrase movement types. Stabler's extension for head movement raises complexity to $\mathcal{O}(n^{4k+12})$ because of the additional spans that are required to keep the head of a constituent available for extraction by head movement. If we interpret $\mathcal{O}(n^4)$ as the price of a single phrase mover, this would mean that the price of having head movement is equivalent to having two phrase movement types.

A more modern parser for MG^S is presented in [7] which lowers the time complexity from Harkema's $\mathcal{O}(n^{4k+4})$ to $\mathcal{O}(n^{2k+3})$. This result is in fact the same algorithm as Harkema's (done through conversion to IRTG) but with more accurate computational analysis. If we apply this revised analysis to Stabler's head movement algorithm we get worst-case parsing complexity of $\mathcal{O}(n^{2k+9})$. Here, the price of each phrase movement is quadratic meaning that head movement now costs the same as having three distinct phrase mover types.

An interesting special type of MG^H is MG_0^H which is an MG that has only head movement without any phrase movement. This MG is more expressive than CFG but less expressive than TAG [21]. Parsing this MG using Stabler's algorithm with new analysis would take $\mathcal{O}(n^9)$ in the worst case. Intuitively, there should be a better algorithm than this because the worst case complexity of TAG, which is more expressive, is just $\mathcal{O}(n^6)$.

This paper presents a new and more efficient algorithm for parsing the full formulation of MG that contains head movement. This greater efficiency is accomplished with a more compact representation of the parse items and inference rules adapted for that compact representation. The parser's worst-case computational complexity is $\mathcal{O}(n^{2k+5})$. In computational terms, this makes parsing head movement as easy as parsing a single phrase movement. In the special case of MG_0^H we get a $\mathcal{O}(n^5)$ parser which is lower than TAG parsing complexity, exactly as we would expect.

2   MG Without Head Movement

A Succinct Minimalist Grammar MG^S [29] is formally defined with a tuple $G = \langle \Sigma, B, Lex, c, \{\text{MERGE,MOVE}\}\rangle$, where Σ is the **vocabulary**, B is a set of **basic features**, Lex is a finite **lexicon** (as defined just below), $c \in B$ is the **start category**, and MERGE and MOVE are the generating functions. The basic features of the set B are concatenated with prefix operators to specify their roles, as follows:

categories, selectees $= B$

$\quad$ **selectors** $= \{=f \mid f \in B\}$

$\quad$ **licensees** $= \{-f \mid f \in B\}$

$\quad$ **licensors** $= \{+f \mid f \in B\}$

Let F be the set of role-marked **features**, that is, the union of the categories, selectors, licensors and licensees. Let $T = \{::,:\}$ be two **types**, indicating "lexical" and "derived" structures, respectively. Let $C = \Sigma^* \times T \times F^*$ be the set of **chains**. Let $E = C^+$ be the set of **expressions**; intuitively, an expression is a chain together with its "moving" sub-chains, if any. All expressions have to respect Shortest Movement Constraint (SMC) which states that no two chains in an expression can have the same initial feature. The functions MERGE and MOVE are defined in Fig. 3. The **lexicon** $Lex \subset \Sigma^* \times \{::\} \times F^*$ is a finite set. The set of all **structures** that can be derived from the lexicon is

$S(G) = \text{closure}(Lex, \{\text{MERGE}, \text{MOVE}\})$. The set of **sentences** $L(G) = \{s \mid s \cdot c \in S(G)$ for some type $\cdot \in \{:, ::\}\}$, where c is the "start" category.

The functions *merge1*, *merge2* *merge3* are special cases of MERGE corresponding respectively to complement merge, specifier merge and merge of an element that is going to move in the future and therefore needs to be kept as a separate chain. Functions *move1* and *move2* are special cases of MOVE corresponding respectively to the movement of an element that is landing and to the movement of an element that will move again in the future. All functions in Fig. 3 have pairwise disjoint domains.

$$\frac{s \; ::= f\gamma \qquad t \cdot f, \alpha_1, \ldots, \alpha_k}{st \; : \; \gamma, \alpha_1, \ldots, \alpha_k}\text{merge1}$$

$$\frac{s \; := f\gamma, \iota_1, \ldots, \iota_l \qquad t \cdot f, \alpha_1, \ldots, \alpha_k}{ts \; : \; \gamma, \iota_1, \ldots, \iota_l, \alpha_1, \ldots, \alpha_k}\text{merge2}$$

$$\frac{s \cdot = f\gamma, \iota_1, \ldots, \iota_l \qquad t \cdot f\delta, \alpha_1, \ldots, \alpha_k}{s \; : \; \gamma, \iota_1, \ldots, \iota_l, t : \delta, \alpha_1, \ldots, \alpha_k}\text{merge3}$$

$$\frac{s \; : \; +f\gamma, \iota_1, \ldots, \iota_l, t : -f, \alpha_1, \ldots, \alpha_k}{ts \; : \; \gamma, \iota_1, \ldots, \iota_l, \alpha_1, \ldots, \alpha_k}\text{move1}$$

$$\frac{s \; : \; +f\gamma, \iota_1, \ldots, \iota_l, t : -f\gamma, \alpha_1, \ldots, \alpha_k}{s \; : \; \gamma, \iota_1, \ldots, \iota_l, t : \gamma, \alpha_1, \ldots, \alpha_k}\text{move2}$$

Fig. 3. Succinct MG. All rules are subject to SMC.

2.1 Parsing with Succinct MG

Most MG parsers are based on the *parsing as deduction* paradigm [24]. In *parsing as deduction* the parser maintains two data structures: a chart and an agenda. These data structures contain *items* that represent a set of derivation trees that share their topmost expression. The chart contains items that are already proved by the parser. The agenda contains items that could in the future combine with items from the chart to prove new items. Parsing starts with the agenda containing all the *axioms* (items that are true without the need for a proof) and an empty chart. When an item is popped out of the agenda, the parser tries to combine it with all the elements in the chart in an attempt to prove new items. For each new item, the parser first checks if it is present in the chart. If it is in the chart, the parser just discards it[1] because that item is already proved. If it is not present in the chart, the parser adds it both to the chart and to the agenda. Parsing stops when either the newly created item is the goal item, in which case

---

[1] Or adds one more backpointer if we want all possible derivation trees instead of a single tree.

parsing is successful, or when the agenda becomes empty, in which case parsing has failed because the sentence is not part of the language.

This description is enough to use this method for recognition. To turn it into a parser, it is sufficient to modify the chart data structure in such a way that each item in it contains a list of backpointers to the items that were used to derive it. When the goal item is constructed, it is enough to follow backpointers to find the full derivation tree.

The first parser for the succinct version of MG was presented by Harkema [10,11]. The items of this parser are equivalent to the MG^S expressions except that instead of strings they contain spans of the strings in the sentence that is being parsed. The axioms of the parser are lexical entries for each word in the sentence with the string replaced by its span in the sentence. Inference rules are exactly the same as in the definition of MERGE and MOVE from Fig. 3.

Harkema's analysis of that algorithm is as follows. The maximal number of items in the chart (its space complexity) is n^{2k+2} because each item contains maximally $k+1$ spans (due to SMC) and each span has 2 indices in range $[0, n]$. In the worst case we will need to pop out n^{2k+2} items from the agenda. The parser needs to check for each of those popped items whether there is a proved item in the chart that could combine with it. In the worst case the number of items in the chart is n^{2k+2}. Therefore the worst-case complexity is $\mathcal{O}(n^{2k+2} \cdot n^{2k+2}) = \mathcal{O}(n^{4k+4})$.

As shown by Fowlie and Koller [7], this analysis was too pessimistic. Through conversion of MG^S to Interpreted Regular Tree Grammars (IRTG) they demonstrated that MG^S can be parsed in $\mathcal{O}(n^{2k+3})$. The same result can be obtained by converting MG^S to MCFG using Michaelis' algorithm [20] and then parsing with some well optimised MCFG parser.

However, conversion to any of these formalisms is not necessary to get efficient MG parsing. It is enough just to implement an efficient lookup in the chart. Optimising for feature lookup is not enough because it will improve only the constants that depend on the grammar. To get asymptotic improvement the lookup needs to be optimised on the item indices instead. For instance, if we have an item that as its main span has $(2, 3)$ and as its initial feature a selector, we know that *merge1* inference rule can combine it only with items whose main span start with 3. If we organise the items in the chart in a way that we can efficiently lookup all items that have particular properties, for instance "all items whose main span starts with 3", then the parsing complexity will be lower. If the popped item has m movers then we know that the item that combines with it certainly does not have more than $k - m$ movers due to SMC constraint. In the case of *merge1* and *merge2* we also know one of the indices of the main span, a fact that reduces the number of possible items that could merge even further. We can calculate complexity by summing over the computations for each possible value of m as follows $\mathcal{O}(\sum_{m=0}^{k} n^{2m+2} \cdot n^{2(k-m)+1}) = \mathcal{O}(n^{2k+3})$. This is a result for *merge1* and *merge2*, which turn out to be more expensive than *merge3*, that does not need to apply concatenation of strings and for that reason can be done in $\mathcal{O}(n^{2k+2})$. Move rules do not need to consult the chart since they are unary rules, but we could still calculate the total number of times they would be applied as $\mathcal{O}(n^{2k+1})$ for *merge1* and $\mathcal{O}(n^{2k+2})$ for *merge2*.

Parsing as deduction systems are essentially logic programs which are evaluated bottom-up (forward chaining) with tabling [6, 15]. For those kinds of programs there is a much simpler way of calculating complexity that is based only on counting unique indices in the antecedents of the most complex inference rule [5, 19]. Clearly in total there cannot be more than k phrasal movers on the antecedent side because the consequent needs to respect SMC. Furthermore, for concatenation rules, i.e. *merge1* and *merge2*, we know that the main spans share at least one index, which leaves us with 3 unique indices for main spans and makes the complexity of those rules $\mathcal{O}(n^{2k+3})$. These rules are also the most complex inference rules and therefore the worst-case complexity of the whole algorithm is $\mathcal{O}(n^{2k+3})$. We will use this method of complexity analysis for parsing algorithms in the rest of the paper.

3 MG with Head Movement

The first description of how to parse MG with head movement was presented by Stabler [27]. It is based on modifying the MG expressions/items in such a way as to split the main string into three sub-parts: the head string s_h, the specifier string s_s and the complement string s_c. The reason for this splitting is to make the head string available for extraction by head movement. MG rules also needed modification to work with this representation. Modifications for *merge2*, *move1* and *move2* are trivial because they cannot trigger head movement. Rules for *merge1* and *merge3* got two additional versions that can trigger head movement and adjoin the complement's head to the selecting head to the left, in case of feature =>f, or to the right, in case of feature <=f.

Stabler's inference rules for MG with head movement[2] are shown in Fig. 4 together with a new complexity analysis for each of the rules calculated by the method of counting indices for each rule. The most complex rule is *merge2* which makes the whole algorithm run in $\mathcal{O}(n^{2k+9})$.

4 Improved Parsing of MG with Head Movement

Stabler's formulation is very compact, but it misses some generalisations that would make the number of indices smaller. For instance, if we take rule merge2 we can see that spans t_s and t_h always share one index because they are concatenated. The same holds for t_h and t_c. This means that the selector needs to visit n^2 constituents in the chart that would produce absolutely the same result

---

[2] Stabler's inference rules had a small mistake for merge3left and merge3right for allowing the possibility of a head constituent being non-lexical. The correct version of inference rules is presented in this paper but it can also be found in [21]. The correction is crucial for the more efficient parsing algorithm.

$$\frac{\epsilon, s, \epsilon ::= f\gamma \qquad t_s, t_h, t_c \cdot f, \alpha_1, \ldots, \alpha_k}{\epsilon, s, t_s t_h t_c : \gamma, \alpha_1, \ldots, \alpha_k} \quad \text{merge1} \quad \mathcal{O}(n^{2k+6})$$

$$\frac{\epsilon, s, \epsilon ::\Leftarrow f\gamma \qquad t_s, t_h, t_c \cdot f, \alpha_1, \ldots, \alpha_k}{\epsilon, s t_h, t_s t_c : \gamma, \alpha_1, \ldots, \alpha_k} \quad \text{merge1right} \quad \mathcal{O}(n^{2k+6})$$

$$\frac{\epsilon, s, \epsilon ::\Rightarrow f\gamma \qquad t_s, t_h, t_c \cdot f, \alpha_1, \ldots, \alpha_k}{\epsilon, t_h s, t_s t_c : \gamma, \alpha_1, \ldots, \alpha_k} \quad \text{merge1left} \quad \mathcal{O}(n^{2k+6})$$

$$\frac{s_s, s_h, s_c := f\gamma, \iota_1, \ldots, \iota_l \qquad t_s, t_h, t_c \cdot f, \alpha_1, \ldots, \alpha_k}{t_s t_h t_c s_s, s_h, s_c : \gamma, \iota_1, \ldots, \iota_l, \alpha_1, \ldots, \alpha_k} \quad \text{merge2} \quad \mathcal{O}(n^{2k+9})$$

$$\frac{s_s, s_h, s_c \cdot = f\gamma, \iota_1, \ldots, \iota_l \qquad t_s, t_h, t_c \cdot f\delta, \alpha_1, \ldots, \alpha_k}{s_s, s_h, s_c : \gamma, \iota_1, \ldots, \iota_l, t_s t_h t_c : \delta, \alpha_1, \ldots, \alpha_k} \quad \text{merge3} \quad \mathcal{O}(n^{2k+8})$$

$$\frac{\epsilon, s, \epsilon ::\Leftarrow f\gamma \qquad t_s, t_h, t_c \cdot f\delta, \alpha_1, \ldots, \alpha_k}{\epsilon, s t_h, \epsilon : \gamma, t_s t_c : \delta, \alpha_1, \ldots, \alpha_k} \quad \text{merge3right} \quad \mathcal{O}(n^{2k+4})$$

$$\frac{\epsilon, s, \epsilon ::\Rightarrow f\gamma \qquad t_s, t_h, t_c \cdot f\delta, \alpha_1, \ldots, \alpha_k}{\epsilon, t_h s, \epsilon : \gamma, t_s t_c : \delta, \alpha_1, \ldots, \alpha_k} \quad \text{merge3left} \quad \mathcal{O}(n^{2k+4})$$

$$\frac{s_s, s_h, s_c : +f\gamma, t : -f, \alpha_1, \ldots, \alpha_k}{t s_s, s_h, s_c : \gamma, \alpha_1, \ldots, \alpha_k} \quad \text{move1} \quad \mathcal{O}(n^{2k+5})$$

$$\frac{s_s, s_h, s_c : +f\gamma, \iota_1, \ldots, \iota_l, t : -f\gamma, \alpha_1, \ldots, \alpha_k}{s_s, s_h, s_c : \gamma, \iota_1, \ldots, \iota_l, t : \gamma, \alpha_1, \ldots, \alpha_k} \quad \text{move2} \quad \mathcal{O}(n^{2k+6})$$

Fig. 4. Stabler's inference rules for MG with head movement together with their computational complexity.

because the two indices that are shared between the components t_s, t_h and t_c are disappearing in the consequent. One could try to reduce this problem by having a unary inference rule that packs all the main components of t before they are combined with merge2 akin to the fold transformation of logic programs [6,14,23]. If we push this operation to its limits we would get $\mathcal{O}(n^{2k+7})$.

To obtain an even lower complexity we need to change the representation of items. Do we need all the three components t_s, t_h and t_c from Stabler's formulation? If we look at the rules in Fig. 4 we can see that when an item is selected by a merge operation, its components are concatenated either as $t_s t_h t_c$ if the head is not extracted or as t_h and $t_s t_c$ if the head is extracted. This highlights a simple, tautological, fact: every head will either be extracted with head movement or it won't. In case it does not participate in head movement, like in the succinct version of MG, there is no need to keep 3 spans to represent projections of that head. It can all be done with a single span and by that reduce the number of indices.

If, on the other hand, the head does participate in head movement then we need only two spans: one for the head s_h and one for the concatenation of specifier and complement s_{sc} because we know with certainty that they will be concatenated after the head is extracted.

If we knew whether the head will move or not the parsing algorithm could be improved significantly, but how can we know if the head will move? The good aspect of chart based parsers is that we do not need to know that ahead of time. We can just encode both variations of the items as axioms and let the parser combine them accordingly. Let us refer to the items whose head *must not* be

$$\frac{s \overset{A}{::} = f\gamma \qquad t \overset{A}{\cdot} f, \alpha_1, \ldots, \alpha_k}{st \overset{A}{\cdot} \gamma, \alpha_1, \ldots, \alpha_k} \qquad \text{merge1A} \qquad \mathcal{O}(n^{2k+3})$$

$$\frac{s, \epsilon \overset{B}{::} = f\gamma \qquad t \overset{A}{\cdot} f, \alpha_1, \ldots, \alpha_k}{s, t \overset{B}{\cdot} \gamma, \alpha_1, \ldots, \alpha_k} \qquad \text{merge1B} \qquad \mathcal{O}(n^{2k+4})$$

$$\frac{s \overset{A}{::} \Leftarrow f\gamma \qquad t_h, t_{sc} \overset{B}{\cdot} f, \alpha_1, \ldots, \alpha_k}{st_h t_{sc} \overset{A}{\cdot} \gamma, \alpha_1, \ldots, \alpha_k} \qquad \text{merge1rightA} \quad \mathcal{O}(n^{2k+4})$$

$$\frac{s, \epsilon \overset{B}{::} \Leftarrow f\gamma \qquad t_h, t_{sc} \overset{B}{\cdot} f, \alpha_1, \ldots, \alpha_k}{st_h, t_{sc} \overset{B}{\cdot} \gamma, \alpha_1, \ldots, \alpha_k} \qquad \text{merge1rightB} \quad \mathcal{O}(n^{2k+5})$$

$$\frac{s \overset{A}{::} \Rightarrow f\gamma \qquad t_h, t_{sc} \overset{B}{\cdot} f, \alpha_1, \ldots, \alpha_k}{t_h s t_{sc} \overset{A}{\cdot} \gamma, \alpha_1, \ldots, \alpha_k} \qquad \text{merge1leftA} \quad \mathcal{O}(n^{2k+4})$$

$$\frac{s, \epsilon \overset{B}{::} \Rightarrow f\gamma \qquad t_h, t_{sc} \overset{B}{\cdot} f, \alpha_1, \ldots, \alpha_k}{t_h s, t_{sc} \overset{B}{\cdot} \gamma, \alpha_1, \ldots, \alpha_k} \qquad \text{merge1leftB} \quad \mathcal{O}(n^{2k+5})$$

$$\frac{s \overset{A}{\cdot} = f\gamma, \iota_1, \ldots, \iota_l \qquad t \overset{A}{\cdot} f, \alpha_1, \ldots, \alpha_k}{ts \overset{A}{\cdot} \gamma, \iota_1, \ldots, \iota_l, \alpha_1, \ldots, \alpha_k} \qquad \text{merge2A} \qquad \mathcal{O}(n^{2k+3})$$

$$\frac{s_h, s_{sc} \overset{B}{\cdot} = f\gamma, \iota_1, \ldots, \iota_l \qquad t \overset{A}{\cdot} f, \alpha_1, \ldots, \alpha_k}{s_h, t s_{sc} \overset{B}{\cdot} \gamma, \iota_1, \ldots, \iota_l, \alpha_1, \ldots, \alpha_k} \qquad \text{merge2B} \qquad \mathcal{O}(n^{2k+5})$$

$$\frac{s \overset{A}{\cdot} = f\gamma, \iota_1, \ldots, \iota_l \qquad t \overset{A}{\cdot} f\delta, \alpha_1, \ldots, \alpha_k}{s \overset{A}{\cdot} \gamma, \iota_1, \ldots, \iota_l, t : \delta, \alpha_1, \ldots, \alpha_k} \qquad \text{merge3A} \qquad \mathcal{O}(n^{2k+2})$$

$$\frac{s_h, s_{sc} \overset{B}{\cdot} = f\gamma, \iota_1, \ldots, \iota_l \qquad t \overset{A}{\cdot} f\delta, \alpha_1, \ldots, \alpha_k}{s_h, s_{sc} \overset{B}{\cdot} \gamma, \iota_1, \ldots, \iota_l, t : \delta, \alpha_1, \ldots, \alpha_k} \qquad \text{merge3B} \qquad \mathcal{O}(n^{2k+4})$$

$$\frac{s \overset{A}{::} \Leftarrow f\gamma \qquad t_h, t_{sc} \overset{B}{\cdot} f\delta, \alpha_1, \ldots, \alpha_k}{st_h \overset{A}{\cdot} \gamma, t_{sc} : \delta, \alpha_1, \ldots, \alpha_k} \qquad \text{merge3rightA} \quad \mathcal{O}(n^{2k+3})$$

$$\frac{s, \epsilon \overset{B}{::} \Leftarrow f\gamma \qquad t_h, t_{sc} \overset{B}{\cdot} f\delta, \alpha_1, \ldots, \alpha_k}{st_h, \epsilon \overset{B}{\cdot} \gamma, t_{sc} : \delta, \alpha_1, \ldots, \alpha_k} \qquad \text{merge3rightB} \quad \mathcal{O}(n^{2k+3})$$

$$\frac{s \overset{A}{::} \Rightarrow f\gamma \qquad t_h, t_{sc} \overset{B}{\cdot} f\delta, \alpha_1, \ldots, \alpha_k}{t_h s \overset{A}{\cdot} \gamma, t_{sc} : \delta, \alpha_1, \ldots, \alpha_k} \qquad \text{merge3leftA} \quad \mathcal{O}(n^{2k+3})$$

$$\frac{s, \epsilon \overset{B}{::} \Rightarrow f\gamma \qquad t_h, t_{sc} \overset{B}{\cdot} f\delta, \alpha_1, \ldots, \alpha_k}{t_h s, \epsilon \overset{B}{\cdot} \gamma, t_{sc} : \delta, \alpha_1, \ldots, \alpha_k} \qquad \text{merge3leftB} \quad \mathcal{O}(n^{2k+3})$$

$$\frac{s \overset{A}{\cdot} + f\gamma, \iota_1, \ldots, \iota_l, t : -f, \alpha_1, \ldots, \alpha_k}{ts \overset{A}{\cdot} \gamma, \iota_1, \ldots, \iota_l, \alpha_1, \ldots, \alpha_k} \qquad \text{move1A} \qquad \mathcal{O}(n^{2k+1})$$

$$\frac{s_h, s_{sc} \overset{B}{\cdot} + f\gamma, \iota_1, \ldots, \iota_l, t : -f, \alpha_1, \ldots, \alpha_k}{s_h, t s_{sc} \overset{B}{\cdot} \gamma, \iota_1, \ldots, \iota_l, \alpha_1, \ldots, \alpha_k} \qquad \text{move1B} \qquad \mathcal{O}(n^{2k+3})$$

$$\frac{s \overset{A}{\cdot} + f\gamma, \iota_1, \ldots, \iota_l, t : -f\gamma, \alpha_1, \ldots, \alpha_k}{s \overset{A}{\cdot} \gamma, \iota_1, \ldots, \iota_l, t : \gamma, \alpha_1, \ldots, \alpha_k} \qquad \text{move2A} \qquad \mathcal{O}(n^{2k+2})$$

$$\frac{s_h, s_{sc} \overset{B}{\cdot} + f\gamma, \iota_1, \ldots, \iota_l, t : -f\gamma, \alpha_1, \ldots, \alpha_k}{s_h, s_{sc} \overset{B}{\cdot} \gamma, \iota_1, \ldots, \iota_l, t : \gamma, \alpha_1, \ldots, \alpha_k} \qquad \text{move2B} \qquad \mathcal{O}(n^{2k+4})$$

Fig. 5. New inference rules for MG with head movement

extracted as items of type A, and items whose head *must* be extracted as items of type B. Type A items will have a single span, just like in succinct MG, while type B items will have two spans: s_h for the head and s_{sc} for specifier-complement. This reduces the space complexity from Stabler's $\mathcal{O}(n^{2k+6})$ to $\mathcal{O}(n^{2k+4})$.

The axioms of the new parser will for each word w_i contain an entry of type A: $w_i \overset{A}{::} \gamma$ and of type B: $w_i, \epsilon \overset{B}{::} \gamma$. These two cover both possible cases of w_i eventually being extracted and not being extracted by the head movement. All MG rules need to be modified accordingly, but the modification is very simple. We have exactly two rules for each rule of Stabler's parser. This is because every rule can have items of type A or type B as its main antecedent item (the item that has selector or licensor as initial feature).

The type of the non-main (selected) item depends on the MG operation: in case we use the head movement we know that the selected item is of type B, otherwise it is of type A. The type of the consequent item is determined by the main antecedent item: the head of the main antecedent item and the head of the consequent is the same and therefore the same constraint on the head movement (whether the head must or must not be extracted) has to stay unchanged.

This gives us the rules of inference shown in Fig. 5. The maximal complexity comes from *merge1rightB*, *merge1leftB* and *merge2B* which make the whole parsing in the worst-case $\mathcal{O}(n^{2k+5})$. This makes the computational price of parsing head movement $\mathcal{O}(n^2)$ which is the same as phrase movement. The number of rules is double the rules of Stabler, but they all have disjoint domains and can still be treated as a two operations MERGE and MOVE. Derivation trees that result from this parsing approach are isomorphic to the derivation trees of Stabler's parser with the only difference in labels of operations containing additional letter A or B.

In case the grammar does not have features for the head movement, we can exclude axioms of type B. This automatically makes parsing MG without head movement $\mathcal{O}(n^{2k+3})$ without doing any transformation to IRTG or MCFG.

5   ATB Head Movement

One interesting variation of head movement is Across-the-Board (ATB) head movement. This variation is not part of Stabler's original formalisation, but is of both theoretical and practical importance. If we accept that in interrogative sentences tense undergoes head movement to adjoin to the complementiser head, then in the case of the coordination of two tense phrases the same tense head has to be simultaneously extracted from both. An example sentence is "Who [does]$_T$ John __ like and Mary __ hate" (which also features ATB phrase movement of *who*).

A formalisation of ATB head movement, as it is used in MGbank, is given in [35]. The inference rules for ATB head movement are similar to ATB Phrase Movement rules from [16]. The formalism of MGbank has special features for coordination that are located on coordinating conjunction and are marked with ^=f. There are two inference rules of relevance here. The first one combines coordinating conjunction with the right conjunct. In the new representation of items the coordinating conjunction will be of type A since its head is not going to undergo head movement, but both the left and the right conjunct will be of type B. The first inference rule has the following form in the more compact representation:

$$\frac{s \overset{A}{::} \ \hat{} =f \ \hat{} =f \ \gamma \qquad t_h,t_{sc} \overset{B}{\vdots} f,\alpha_1,\ldots,\alpha_k}{t_h,st_{sc} \overset{B}{\vdots} \ \hat{} =f \ \gamma,\alpha_1,\ldots,\alpha_k}$$

The second inference rule combines the result of the first rule with the left conjunct. Because this is ATB head movement we know that heads and all the moving chains have to unify between two antecedents. That gives us the following inference rule.

$$\frac{t_h,s_{sc} \overset{B}{\vdots} f,\alpha_1,\ldots,\alpha_k \qquad t_h,t_{sc} \overset{B}{\vdots} \ \hat{} =f \ \gamma,\alpha_1,\ldots,\alpha_k}{t_h,s_{sc}t_{sc} \overset{B}{\vdots} \gamma,\alpha_1,\ldots,\alpha_k}$$

Even though the last rule uses two items of type B, the complexity is still low $\mathcal{O}(n^{2k+5})$ because the two antecedents share the same head.

6   Affix Hopping

Affix hopping [3] is a morphosyntactic operation similar to head movement and in some sense its opposite. Affix hopping can be interpreted as a downward head movement where the head of the selector is moving to adjoin to the complement's head. The main motivation for this rule, as is apparent from its name, is to move the tense affix to the verb in languages like English where V-to-T head movement cannot occur. For instance, in "John __ really like $[-s]_T$ Mary" the tense affix "$-s$" is often assumed to have undergone the affix hopping to adjoin to the main verb stem "like".

$$\frac{\epsilon,s,\epsilon \ :: \approx\!\!> f\gamma \qquad t_s,t_h,t_c \ \cdot \ f,\alpha_1,\ldots,\alpha_k}{\epsilon,\epsilon,t_st_hst_c \ : \ \gamma,\alpha_1,\ldots,\alpha_k} \quad \text{merge1HopRight} \quad \mathcal{O}(n^{2k+5})$$

$$\frac{\epsilon,s,\epsilon \ :: <\!\!\approx f\gamma \qquad t_s,t_h,t_c \ \cdot \ f,\alpha_1,\ldots,\alpha_k}{\epsilon,\epsilon,t_sst_ht_c \ : \ \gamma,\alpha_1,\ldots,\alpha_k} \quad \text{merge1HopLeft} \quad \mathcal{O}(n^{2k+5})$$

$$\frac{\epsilon,s,\epsilon \ :: \approx\!\!> f\gamma \qquad t_s,t_h,t_c \ \cdot \ f\delta,\alpha_1,\ldots,\alpha_k}{\epsilon,\epsilon,\epsilon \ : \ \gamma,t_st_hst_c:\delta,\alpha_1,\ldots,\alpha_k} \quad \text{merge3HopRight} \quad \mathcal{O}(n^{2k+3})$$

$$\frac{\epsilon,s,\epsilon \ :: <\!\!\approx f\gamma \qquad t_s,t_h,t_c \ \cdot \ f\delta,\alpha_1,\ldots,\alpha_k}{\epsilon,\epsilon,\epsilon \ : \ \gamma,t_sst_ht_c:\delta,\alpha_1,\ldots,\alpha_k} \quad \text{merge3HopLeft} \quad \mathcal{O}(n^{2k+3})$$

Fig. 6. Stabler's inference rules for MG with affix hopping.

Affix hopping was formalised in Stabler's paper on head movement [27] with the inference rules shown in Fig. 6 where a special selector feature $\approx\!\!> f$ or $<\!\!\approx f$ is used to trigger affix hopping. Owing to our adoption of a different item representation to get a faster head movement parser, these affix hopping rules cannot be directly supported. For head movement we have exploited the fact that the main strings of every item are concatenated either as t_h, t_{sc} or t_{shc}. However, affix hopping has additional string combinations like t_{hc} and t_{sh}.

To account for those additional options we create two more types of items. Items of type C are items that must adjoin some head via affix hopping to the right of the head and that is why they have two main strings: a specifier-head string t_{sh} and a complement string t_c. Items of type D are similar except that they account for affix hopping to the right of the head. Items of type C and D do not trigger affix hopping but only accept (host) affix that has hopped. The new inference rules for affix hopping are shown in Fig. 7.

$$\frac{s \overset{A}{::} \approx\!\!> f\gamma \qquad t_{sh}, t_c \overset{C}{\cdot} f, \alpha_1, \ldots, \alpha_k}{t_{sh}st_c \overset{A}{:} \gamma, \alpha_1, \ldots, \alpha_k} \qquad \text{merge1HopRight} \qquad \mathcal{O}(n^{2k+4})$$

$$\frac{s \overset{A}{::} <\!\!\approx f\gamma \qquad t_s, t_{hc} \overset{D}{\cdot} f, \alpha_1, \ldots, \alpha_k}{t_sst_{hc} \overset{A}{:} \gamma, \alpha_1, \ldots, \alpha_k} \qquad \text{merge1HopLeft} \qquad \mathcal{O}(n^{2k+4})$$

$$\frac{s \overset{A}{::} \approx\!\!> f\gamma \qquad t_{sh}, t_c \overset{C}{\cdot} f\delta, \alpha_1, \ldots, \alpha_k}{\epsilon \overset{A}{:} \gamma, t_{sh}st_c : \delta, \alpha_1, \ldots, \alpha_k} \qquad \text{merge3HopRight} \qquad \mathcal{O}(n^{2k+2})$$

$$\frac{s \overset{A}{::} <\!\!\approx f\gamma \qquad t_s, t_{hc} \overset{D}{\cdot} f\delta, \alpha_1, \ldots, \alpha_k}{\epsilon \overset{A}{:} \gamma, t_sst_{hc} : \delta, \alpha_1, \ldots, \alpha_k} \qquad \text{merge3HopLeft} \qquad \mathcal{O}(n^{2k+2})$$

Fig. 7. Improved inference rules for MG with affix hopping.

We allow the selector constituent to be only of type A because its head will not be able to undergo head movement later. This is a stricter definition than Stabler's because the latter allows the empty string (which is not a head of the new constituent but only a replacement for the real head) to participate in head movement. This modification does influence the set of the derivations that could be built by the parser but in a good way: it does not make sense for an affix to undergo affix hopping downwards and then head movement upwards (or fake head movement of an empty string).

Similarly, we do not allow the selecting item to be of type C or D because after the affix has hopped, its slot will be empty, so it does not make sense for another affix to hop to its original place. Affixes move to attach to some overt word and there is none at this slot.

The rules in Fig. 7 show how items of type C and D are used for affix hopping. But it still remains to show how items of type C and D are built. To build items of type C and D, we need to add them to the agenda as axioms for all lexical items (just as was done for items of type B) and to use additional inference rules that are variations of rules for items of type A and B from Fig. 5. The additional rules for items of type C are shown in Fig. 8. Similar rules are trivial to make for items of type D.

$$\frac{s,\epsilon \overset{C}{::=} f\gamma \qquad t \overset{A}{\cdot} f,\alpha_1,\ldots,\alpha_k}{s,t \overset{C}{:} \gamma,\alpha_1,\ldots,\alpha_k} \qquad \text{merge1C} \qquad \mathcal{O}(n^{2k+4})$$

$$\frac{s,\epsilon \overset{C}{::} \Leftarrow f\gamma \qquad t_h,t_{sc} \overset{B}{\cdot} f,\alpha_1,\ldots,\alpha_k}{st_h,t_{sc} \overset{C}{:} \gamma,\alpha_1,\ldots,\alpha_k} \qquad \text{merge1rightC} \quad \mathcal{O}(n^{2k+5})$$

$$\frac{s,\epsilon \overset{C}{::} \Rightarrow f\gamma \qquad t_h,t_{sc} \overset{B}{\cdot} f,\alpha_1,\ldots,\alpha_k}{t_h s,t_{sc} \overset{C}{:} \gamma,\alpha_1,\ldots,\alpha_k} \qquad \text{merge1leftC} \quad \mathcal{O}(n^{2k+5})$$

$$\frac{s_{sh},s_c \overset{C}{:} = f\gamma,\iota_1,\ldots,\iota_l \qquad t \overset{A}{\cdot} f,\alpha_1,\ldots,\alpha_k}{t s_{sh},s_c \overset{C}{:} \gamma,\iota_1,\ldots,\iota_l,\alpha_1,\ldots,\alpha_k} \qquad \text{merge2C} \qquad \mathcal{O}(n^{2k+5})$$

$$\frac{s_{sh},s_c \overset{C}{\cdot} = f\gamma,\iota_1,\ldots,\iota_l \qquad t \overset{A}{\cdot} f\delta,\alpha_1,\ldots,\alpha_k}{s_{sh},s_c \overset{C}{:} \gamma,\iota_1,\ldots,\iota_l,t:\delta,\alpha_1,\ldots,\alpha_k} \qquad \text{merge3C} \qquad \mathcal{O}(n^{2k+4})$$

$$\frac{s,\epsilon \overset{C}{::} \Leftarrow f\gamma \qquad t_h,t_{sc} \overset{B}{\cdot} f\delta,\alpha_1,\ldots,\alpha_k}{st_h,\epsilon \overset{C}{:} \gamma,t_{sc}:\delta,\alpha_1,\ldots,\alpha_k} \qquad \text{merge3rightC} \quad \mathcal{O}(n^{2k+3})$$

$$\frac{s,\epsilon \overset{C}{::} \Rightarrow f\gamma \qquad t_h,t_{sc} \overset{B}{\cdot} f\delta,\alpha_1,\ldots,\alpha_k}{t_h s,\epsilon \overset{C}{:} \gamma,t_{sc}:\delta,\alpha_1,\ldots,\alpha_k} \qquad \text{merge3leftC} \quad \mathcal{O}(n^{2k+3})$$

$$\frac{s_{sh},s_c \overset{C}{:} +f\gamma,\iota_1,\ldots,\iota_l,t:-f,\alpha_1,\ldots,\alpha_k}{t s_{sh},s_c \overset{C}{:} \gamma,\iota_1,\ldots,\iota_l,\alpha_1,\ldots,\alpha_k} \qquad \text{move1C} \qquad \mathcal{O}(n^{2k+3})$$

$$\frac{s_{sh},s_c \overset{C}{:} +f\gamma,\iota_1,\ldots,\iota_l,t:-f\gamma,\alpha_1,\ldots,\alpha_k}{s_{sh},s_c \overset{C}{:} \gamma,\iota_1,\ldots,\iota_l,t:\gamma,\alpha_1,\ldots,\alpha_k} \qquad \text{move2C} \qquad \mathcal{O}(n^{2k+4})$$

Fig. 8. Additional inference rules for building items of type C that must host affix hopping in the later part of the derivation.

The rules merge1leftC and merge3leftC may at first appear somewhat surprising. They move the head of the complement to the left of the specifier-head string making a complex head-specifier-head. This may appear to break the rules of head movement which state that the head adjoins to another head and there cannot be any phrase in between them. However, this is not a problem in this case. Since we know that head movement can be triggered only by a lexical item we can be certain that there is no specifier in its specifier-head string, so the final result of the concatenation is a head-head complex.

7   Conclusion

The main motivation for this parser is lowering the worst-case complexity of parsing MG that contains head movement, ATB head movement and affix hopping. Given the recent appearance of the new dataset with MG derivation trees [34] which contains head movement in every derivation, this algorithm is likely to be not only of theoretical but also of practical significance.

MGbank has lead to the first wide coverage Minimalist parser [36]. This is a neural network based parser that uses A* search with Harkema's inference rules and Stabler's approach to head movement. A* improves the best and average-case scenario but the worst-case stays the same $\mathcal{O}(n^{4k+12})$ which is $\mathcal{O}(n^{28})$ for the MGbank grammar. With the algorithm proposed in this paper the overall worst-case complexity will be reduced to $\mathcal{O}(n^{13})$. More importantly, it will potentially also improve average-case complexity because of the more optimal lookup.

The complexity of the parser presented here is $\mathcal{O}(n^{2k+5})$ which is just $\mathcal{O}(n)$ bigger than its space complexity. Further asymptotic improvements would probably require either finding some more compact item representation or abandoning the *parsing as deduction* approach and using some alternative approach akin to Valiant style parsing.

Acknowledgement. I am grateful to John Torr and the three anonymous reviewers for comments that have greatly improved this paper. This work was supported by ERC H2020 Advanced Fellowship GA 742137 SEMANTAX grant.

References

1. Adger, D.: Core Syntax: A Minimalist Approach, vol. 33. Oxford University Press, Oxford (2003)
2. Carnie, A.: Syntax: A Generative Introduction, 3rd edn. Wiley, Hoboken (2013)
3. Chomsky, N.: Syntactic Structures. Mouton, The Hague (1957)
4. Chomsky, N.: The Minimalist Program. MIT Press, Cambridge (1995)
5. Dowling, W.F., Gallier, J.H.: Linear-time algorithms for testing the satisfiability of propositional horn formulae. J. Logic Program. **1**(3), 267–284 (1984)
6. Eisner, J., Blatz, J.: Program transformations for optimization of parsing algorithms and other weighted logic programs. In: Wintner, S. (ed.) Proceedings of the 11th Conference on Formal Grammar 2006, pp. 45–85. CSLI Publications, Stanford (2007)
7. Fowlie, M., Koller, A.: Parsing minimalist languages with interpreted regular tree grammars. In: Proceedings of the 13th International Workshop on Tree Adjoining Grammars and Related Formalisms, pp. 11–20 (2017)
8. Gerth, S.: Memory Limitations in Sentence Comprehension: A Structural-based Complexity Metric of Processing Difficulty, vol. 6. Universitätsverlag Potsdam, Potsdam (2015)
9. Graf, T., Monette, J., Zhang, C.: Relative clauses as a benchmark for minimalist parsing. J. Lang. Model. **5**(1), 57–106 (2017)
10. Harkema, H.: A recognizer for minimalist grammars. In: Proceedings of the Sixth International Workshop on Parsing Technologies (IWPT 2000), pp. 111–122. Springer, Berlin (2000)
11. Harkema, H.: A recognizer for minimalist languages. In: Bunt, H., Carroll, J., Satta, G. (eds.) New Developments in Parsing Technology. Text, Speech and Language Technology, vol. 23, pp. 251–268. Springer, Dordrecht (2005). https://doi.org/10. 1007/1-4020-2295-6_12
12. Hunter, T.: Left-corner parsing of minimalist grammars. Technical report, UCLA (2017). forthcoming

13. Hunter, T., Stanojević, M., Stabler, E.: The active-filler strategy in a move-eager left-corner minimalist grammar parser. In: Proceedings of the 9th Workshop on Cognitive Modeling and Computational Linguistics (CMCL 2019). Association for Computational Linguistics (2019)
14. Johnson, M.: Transforming projective bilexical dependency grammars into efficiently-parsable CFGs with unfold-fold. In: Proceedings of the 45th Annual Meeting of the Association of Computational Linguistics, pp. 168–175. Association for Computational Linguistics (2007)
15. Kanazawa, M.: Parsing and generation as datalog queries. In: Proceedings of the 45th Annual Meeting of the Association of Computational Linguistics, pp. 176–183. Association for Computational Linguistics, Prague, June 2007
16. Kobele, G.M.: Across-the-board extraction in minimalist grammars. In: Proceedings of the Ninth International Workshop on Tree Adjoining Grammar and Related Frameworks (TAG+ 9), pp. 113–120. Association for Computational Linguistics (2008)
17. Kobele, G.M., Gerth, S., Hale, J.: Memory resource allocation in top-down minimalist parsing. In: Morrill, G., Nederhof, M.J. (eds.) FG 2012-2013. LNCS, vol. 8036, pp. 32–51. Springer, Heidelberg (2013). https://doi.org/10.1007/978-3-642-39998-5_3
18. Koopman, H., Szabolcsi, A.: Verbal Complexes. MIT Press, Cambridge (2000)
19. McAllester, D.: On the complexity analysis of static analyses. J. ACM **49**(4), 512–537 (2002)
20. Michaelis, J.: Derivational minimalism is mildly context–sensitive. In: Moortgat, M. (ed.) LACL 1998. LNCS (LNAI), vol. 2014, pp. 179–198. Springer, Heidelberg (2001). https://doi.org/10.1007/3-540-45738-0_11
21. Michaelis, J.: Notes on the complexity of complex heads in a minimalist grammar. In: Proceedings of the Sixth International Workshop on Tree Adjoining Grammar and Related Frameworks (TAG+ 6), pp. 57–65 (2002)
22. Radford, A.: Minimalist Syntax: Exploring the Structure of English. Cambridge University Press, Cambridge (2004)
23. Shepherdson, J.C.: Unfold/fold transformations of logic programs. Math. Struct. Comput. Sci. **2**(2), 143–157 (1992)
24. Shieber, S.M., Schabes, Y., Pereira, F.C.: Principles and implementation of deductive parsing. J. Logic Program. **24**(1–2), 3–36 (1995)
25. Sportiche, D., Koopman, H., Stabler, E.: An Introduction to Syntactic Analysis and Theory. Wiley, Hoboken (2013)
26. Stabler, E.: Derivational minimalism. In: Retoré, C. (ed.) LACL 1996. LNCS, vol. 1328, pp. 68–95. Springer, Heidelberg (1997). https://doi.org/10.1007/BFb0052152
27. Stabler, E.P.: Recognizing head movement. In: de Groote, P., Morrill, G., Retoré, C. (eds.) LACL 2001. LNCS (LNAI), vol. 2099, pp. 245–260. Springer, Heidelberg (2001). https://doi.org/10.1007/3-540-48199-0_15
28. Stabler, E.: Comparing 3 perspectives on head movement. In: From Head Movement and Syntactic Theory, UCLA/Potsdam Working Papers in Linguistics, pp. 178–198 (2003)
29. Stabler, E.: Computational perspectives on minimalism. In: Boeckx, C. (ed.) Oxford Handbook of Linguistic Minimalism, pp. 617–641. Oxford University Press, Oxford (2011)
30. Stabler, E.: Top-down recognizers for MCFGs and MGs. In: Proceedings of the 2nd Workshop on Cognitive Modeling and Computational Linguistics, pp. 39–48. Association for Computational Linguistics, Portland, Oregon, June 2011

31. Stabler, E.: Two models of minimalist, incremental syntactic analysis. Top. Cogn. Sci. **5**(3), 611–633 (2013)
32. Stanojević, M., Stabler, E.: A sound and complete left-corner parsing for minimalist grammars. In: Proceedings of the Eight Workshop on Cognitive Aspects of Computational Language Learning and Processing, pp. 65–74. Association for Computational Linguistics (2018)
33. Stanojević, M.: Minimalist grammar transition-based parsing. In: Amblard, M., de Groote, P., Pogodalla, S., Retoré, C. (eds.) LACL 2016. LNCS, vol. 10054, pp. 273–290. Springer, Heidelberg (2016). https://doi.org/10.1007/978-3-662-53826-5_17
34. Torr, J.: Constraining MGbank: agreement, l-selection and supertagging in minimalist grammars. In: Proceedings of the 56th Annual Meeting on Association for Computational Linguistics. Association for Computational Linguistics, Melbourne (2018)
35. Torr, J., Stabler, E.: Coordination in minimalist grammars. In: Proceedings of the 12th Annual Workshop on Tree-Adjoining Grammars and Related Formalisms (TAG+ 12) (2016)
36. Torr, J., Stanojević, M., Steedman, M., Cohen, S.: Wide-coverage neural A* parsing for minimalist grammars. In: Proceedings of the 56th Annual Meeting of the Association for Computational Linguistics (Vol. 1: Long Papers). Association for Computational Linguistics, Florence (2019)

Author Index

Printed in the United States
By Bookmasters